Greening of Industry Networks Studies

Volume 6

Series editors
Diego A. Vazquez-Brust, Egham, Surrey, UK
Joseph Sarkis, Worcester, MA, USA

More information about this series at http://www.springer.com/series/10444

María-Laura Franco-García
Jorge Carlos Carpio-Aguilar • Hans Bressers
Editors

Towards Zero Waste

Circular Economy Boost, Waste to Resources

Springer

Editors
María-Laura Franco-García
University of Twente
Enschede, The Netherlands

Jorge Carlos Carpio-Aguilar
Walmart Mexico
CDMX, Mexico

Hans Bressers
University of Twente
Enschede, The Netherlands

ISSN 2543-0246 ISSN 2543-0254 (eBook)
Greening of Industry Networks Studies
ISBN 978-3-030-06550-8 ISBN 978-3-319-92931-6 (eBook)
https://doi.org/10.1007/978-3-319-92931-6

Softcover re-print of the Hardcover 1st edition 2019

This Springer imprint is published by the registered company Springer Nature Switzerland AG
The registered company address is: Gewerbestrasse 11, 6330 Cham, Switzerland

Contents

Contributors

Oekan S. Abdoellah Graduate School on Environmental Sciences, Universitas Padjadjaran, Bandung, West Java Province, Indonesia

Hans Bressers University of Twente, Enschede, The Netherlands

Jaime Cámara-Creixell PetStar, Toluca de Lerdo, Mexico

Jorge Carlos Carpio-Aguilar Walmart Mexico, CDMX, Mexico

José-Luis Cruz-Pastrana Instituto Tecnológico de Estudios Superiores Monterrey, Cd López Mateos, Estado de México, México

Luz del Carmen Díaz-Peña Instituto Tecnológico y de Estudios Superiores de Monterrey, Puebla, Mexico

Hans Dieleman Urban Studies Center, Autonomous University of Mexico City, México City, Mexico

María-Laura Franco-García University of Twente, Enschede, The Netherlands

Maria Gavrilescu Faculty of Chemical Engineering and Environmental Protection, Department of Environmental Engineering and Management, "Gheorghe Asachi" Technical University of Iasi, Iasi, Romania

Academy of Romanian Scientists, Bucharest, Romania

Cristina Ghinea Faculty of Food Engineering, Stefan cel Mare University of Suceava, Suceava, Romania

Marco Antonio González-Pérez Faculty for Higher Education (FES), Iztacala Autonomous National University of Mexico (UNAM)|, Tlalnepantla, Mexico State, Mexico

Willem Haanstra Faculty of Engeneering Technology, Design, Production and Management, University of Twente, Enschede, The Netherlands

Mauricio Hernández-Sarabia Faculty of Engineering, Universidad de Ibagué, Ibagué, Colombia

Yuli Astuti Hidayati Graduate School on Environmental Sciences, Universitas Padjadjaran, Bandung, West Java Province, Indonesia

Sharon Hophmayer-Tokich CSTM Department of Governance and Technology for Sustainability, University of Twente, Enschede, The Netherlands

Denny Kurniadie Graduate School on Environmental Sciences, Universitas Padjadjaran, Bandung, West Java Province, Indonesia

María-Concepción Martínez-Rodríguez Centro Interdisciplinario de Investigaciones y Estudios sobre Medio Ambiente y Desarrollo, Instituto Politécnico Nacional (IPN-CIIEMAD), Unidad Zacatenco, México City, Mexico

Juli Nurdiana Faculty of Engineering, Mulawarman University, Samarinda, Indonesia

Vivian Plasencia-Vélez World Resources Institute, Belisario Domínguez 8, Col. Villa Coyoacán, Delegación Coyoacán, Mexico City, Mexico

John Rincón-Moreno University of Twente, alumni, Enschede, The Netherlands

Boelo Schuur Faculty of Science and Technology, Sustainable Process Technology, University of Twente, Enschede, The Netherlands

Carlos Sheel-Mayenberger Instituto Tecnológico de Estudios Superiores de Monterrey, EGADE Business School, Monterrey, NL, Mexico

Vina Septi Suherman Doctoral Program in Environmental Sciences, Universitas Padjadjaran, Bandung, West Java Province, Indonesia

Miguel Angel Tinoco-Castrejón Instituto Tecnológico y de Estudios Superiores de Monterrey, Puebla, Mexico

Marten Toxopeus Faculty of Engeneering Technology, Design, Production and Management, University of Twente, Enschede, The Netherlands

Zongguo Wen Environmental Planning and Management, Tsinghua University, Beijing, People's Republic of China

Research Center of Circular Economy, School of Environment, Tsinghua University, Beijing, People's Republic of China

Yanyan Xue Research Center of Circular Economy, School of Environment, Tsinghua University, Beijing, People's Republic of China

List of Figures

List of Tables

Chapter 1
Towards Zero Waste, Circular Economy Boost: Waste to Resources

María-Laura Franco-García, Jorge Carlos Carpio-Aguilar, and Hans Bressers

Abstract This chapter provides an overview of *circular economy* (CE) and related concepts such as eco-design, biomimicry and eco-industrial development, exploring theoretical and empirical overlaps and complementarities. The circular economy concept is regarded in this chapter as the inspiration to guide public, civil societal and private sectors towards zero waste practices. Firstly, these concepts are independently reviewed from existing literature, categorising some of their nexus. Secondly, this chapter explores the distinction and tensions between the zero waste concept and the zero waste to landfill concept.

Acknowledging how the nature of the stakeholders influences their resource management, the remaining chapters of this book were grouped per type of the main stakeholder (public and private sector and civil society) that promotes CE practices.

Keywords Circular economy · Zero waste · Changing wastes into resources

1.1 Sustainability and Circular Economy-Related Concepts

Resources scarcity and pollution of resources represent the driver of many initiatives to develop and implement production-consumption innovations that decouple economic growth from resource depletion and emissions of pollutants. In the same line of thoughts, the concept of *circular economy* (CE) represents one of the most mentioned frameworks trying to integrate economic activity and environmental wellbeing in a sustainable way (Murray et al. 2015). In a large number of studies, economic sustainability has taken advantaged attention over environmental and

M.-L. Franco-García (✉) · H. Bressers
University of Twente, Enschede, The Netherlands
e-mail: m.l.francogarcia@utwente.nl; j.t.a.bressers@utwente.nl

J. C. Carpio-Aguilar
Walmart Mexico, CDMX, Mexico
e-mail: jorge.carpio@walmart.com

M.-L. Franco-García et al. (eds.), *Towards Zero Waste*, Greening of Industry Networks Studies 6, https://doi.org/10.1007/978-3-319-92931-6_1

social aspects (Schneider 2014), but it is also true that CE has brought a systematic approach to incorporate environmental concerns to economic development. More importantly, CE has inspired alternative business models that allow to re-examine it as the new approach to sustainability, understanding the sustainability concept as the equilibrium state of economic, environmental and social concerns of the society (Dresner 2002).

Despite the growth of CE as a business paradigm towards sustainability, the concept has been explored at different levels but no so much debated academically within business and sustainability literature (Murray et al. 2015). In particular, the term sustainability cannot fully be associated with CE because the social component of sustainability is practically dismissed (Schneider 2014) which, on the other hand, represents an opportunity for policymakers and the business community. In fact, the theoretical frameworks in association with CE bond to environmental economics, ecologic economics and industrial ecology theories reflecting the nature of its current developments and direction.

From the perspective of environmental-economic linkages, the foundation of the CE paradigm has a relative long previous story which embeds, simultaneously through different levels, the principles of Lansink's waste hierarchical management (2010), eco-design (Brezet and Hemel 1997), cradle to cradle (McDonough and Braungart 2002), biomimicry (Benyus 1997), and industrial ecology (Frosch 1992; Graedel 1996) which are among the most reported in literature. All of them are incorporated, to some extent, to frameworks promoted by think-tanks such as the Ellen MacArthur Foundation and consultant agencies (McKinsey & Company, Accenture, etc.). Some of them captured our attention here due to their clear contribution to the discussions on the nexus between CE, zero waste to landfill and the shift of wastes to resources.

Clearly, Lansink's waste hierarchical management (WHM) bridges the three terms because he prioritises the operations of reduce, reuse and recycle (3Rs) over incineration and landfilling, trying to stress the prevention of waste generation that requires any specific management (treatment) or disposal. In fact, the WHM became very influential across countries transforming the waste policy in many of them. For instance, in the Netherlands, the operationalisation of the WHM in the 1990s shaped 30 ambitious waste stream programmes that included batteries, tires, and packaging, among others. These programmes not only improved the waste management at large but also created "new" types of businesses to transport and treat the separated waste streams.

In parallel with the prevention and recycling policies, eco-design was also promoted in the 1990s to be coupled with the recycled resources which were released from the 30 waste stream programmes. This enlarged the scope of environmental protection approaches because the eco-design philosophy promotes a more environmentally conscious product. More importantly, under the eco-design philosophy, in every single stage of the life cycle of the product, environmental concerns should steer the design processes and decisions. Nowadays, the eco-design concept inspires businesses to improve the environmental performance of their products by integrating reuse and recycling of some of the product components into their designs. Even further, eco-design was a pioneer in the adoption of innovative approaches, such as cradle to cradle (McDonough and Braungart 2002) which brought a new environmental-economic perspective: "wastes are food!" This was a crucial milestone towards the transition from waste management to resource management. Without doubt, resource management, as much as the CE paradigm, requires the integration of efforts aligned to close resources' (materials') loops. Closing the materials' loops implies the reduction of raw materials extraction and the decline of waste disposal, simultaneously. But the CE paradigm entails much more than "only" closing the materials' loops as it is stated by the Ellen MacArthur Foundation (2013) to guide on its adoption.

Ellen MacArthur published in 2013 four categories of actions that were introduced into their framework: (1) circular product design and production (eco-design methods aiming to enable product reuse, refurbishment and recycling and designing with less hazardous substances), (2) business models (diffusion of new models, such as service systems, rather than product ownership), (3) cascade/reverse skills (supporting closed-loop cycles, e.g. with innovative technologies for high-quality recycling or for cascading use of materials), and (4) cross-cycle and cross-sector collaboration (prevention of by-products to become waste through an effective industrial symbiosis). It is worth noting that industrial symbiosis (Yuang and Shi 2009) occurs at the interfirm level because it includes exchange options among several organisations. The expression "symbiosis" builds on the notion of biological symbiotic relationships in nature, in which at least two otherwise unrelated species exchange materials, energy or information in a mutually beneficial manner. The symbioses need not occur within the strict boundaries of a "park" (Chertow 2000).

The eco-industrial development theory also mentions closed loops of material flow within the broader economic system through mimicking the natural ecosystem metaphor as the baseline of CE (Geng and Doberstein 2008). Even further, cradle to cradle (C2C) avails itself of biomimetics, which "tries to translate natural production processes to man-made industrial production processes with the eventual goal of harmonising humans and their environment by creating sustainable production processes which then lead to a healthy green environment" (Benyus 1997). It is interesting to observe the interlinkages between closed loops of material flows, biomimicry, C2C and industrial ecology. In particular, when this industrial ecology can be thought of and approached in much the same way as a biological ecosystem (Chertow 2000) and that in its ideal form, it would strive towards integration of activities and cyclisation of resources, as do natural ecosystems. Moreover and now in relation to CE, Murray et al. (2015) suggested that circular economy has been framed in an almost identical way as industrial ecology, with three levels of initiatives: (a) single enterprise, involving a firm-level study of cleaner production, such as the work of Yuan and Shi (2009) on eco-industrial initiatives at a smelter (this level can be comparable with industrial symbiosis); (b) interfirm clusters at supply chain level, represented by eco-industrial parks (EIPs) and involving industrial symbiosis; and (c) entire cities/municipalities, incorporating industrial metabolism (Chertow and Lombardi 2005; L. Zhang et al. 2010).

In addition to biomimicry approaches, the C2C framework distinguishes a technical cycle from a biological cycle of a production chain. The technical cycle consists of technical nutrients, which are inorganic, man-made materials (e.g. plastics or metals) that can be reused over and over again, hence creating a continuous utilisation cycle. Recycling thereby encompasses recycling without a change in quality as well as upcycling and down-cycling processes, respectively. The biological cycle contains biological nutrients, which are organic materials that are disposable without any negative consequences for the environment.

At present, some companies are claiming to launch "circular products". According to Ellen MacArthur Foundation (2013), "circular products are the result of a design that can significantly reduce the material bill and the expense of waste disposal". Those products can create new enterprises that meet needs of new customers as the population grows". It is argued that not only circular products are produced *resource-wise effectively* but also *respectfully*[1] towards its stakeholders' wellbeing. Respectfulness refers to a shift towards more responsible and holistic business models that is increasingly observed with companies dedicating their purpose to solving social and environmental problems while being profitable businesses (Haugh 2005).

The terms "effectiveness" and "respect" need to be tangible; hence, measurement tools to account the environmental and social impacts of the CE implementation are relevant to distinguish their advantages when comparing with "non-circular" products. There are several tools aiming to increase the products' accountability, and one of the most mature is the life cycle assessment (LCA) method which traditionally is meant for assessing (and predicting) environmental effects. Currently, there are some efforts to measure the social aspects under the LCA rationale. In fact, LCA offers the perfect setting to include all relevant stakeholders along the value chain of a product, as it intrinsically encompasses the entire product life cycle.

1.2 Zero Waste Concept vs Zero Waste to Landfill

In this section, we elaborate on two concepts that are often discussed in line with the CE concept: zero waste (ZW) and zero waste to landfill (ZWTL). The ZW concept and principles have been promoted by the Zero Waste International Alliance (ZWIA) who also has developed related policies with the goal to create awareness in the society and businesses about the benefits gained when the waste is regarded as a *resource* (ZWIA 2015). Under the ZW context, it is important for international, national or local recognition to differentiate what ZWTL implies since this is actually based on one of the zero waste business principles. From practitioners'

[1] A shift towards more responsible and holistic business models has been observed with companies dedicating their purpose to solving social and environmental problems while being profitable businesses (Haugh 2005).

viewpoint, ZWTL is one of those innovative models that are considered disruptive in terms of the use of resources in all its value chain.

The ZWIA defined since 2009 "Zero Waste as a goal that is ethical, economical, efficient and visionary, to guide people in changing their lifestyles and practices to emulate sustainable natural cycles, where all discarded materials are designed to become resources for others to use", similar goals than CE. "Zero waste means designing and managing products and processes to systematically avoid and eliminate the volume and toxicity of waste and materials, conserve and recover all resources and not burn or bury them. Ideally, implementing zero waste will eliminate all discharges to land, water or air that are a threat to planetary, human, animal or plant health" (ZWIA 2015).

From the Citizen's Agenda 2001 for Zero Waste, the definition of "zero waste" can be inferred, which is also considered as a complement to the community and industry positions that combine ethical practices and an economic vision, recognising recycling and its limitations in order to reconfigure the one-way industry system into a circular economy (Connett and Sheehan 2001). From the same perspective of Connett and Sheehan, the combination of community practices such as reuse, repair and recycle and the industry practices in order to achieve ZW is promoting the need to develop both sustainable communities and companies.

Zero waste is also considered as an alternative solution for waste management problems. This concept can stimulate sustainable production and consumption, recovery and recycling and restricts incineration and landfilling (Zaman 2015). During the discussions on circular economy, it is mentioned that "a way to close the loop is to ban landfilling" (Zero Waste Europe (ZWE) 2015). But, in reality, what happens is that there is an increase of waste incineration rates; see the cases of Denmark, Germany, the Netherlands, Norway, Sweden, Austria and Switzerland. And even though it substantially reduces the waste volume, incineration technology generates other types of waste (fly and bottom ashes) (ZWE 2015). According to ZWE (2015), the concept of "zero waste to landfill" or landfill ban has contributed to the linear economy because it does not incentive the prevention of waste management.

In support of the above, Zaman in 2015 described the ZW concept as an alternative solution for waste management problems at different phases of production and consumption. He favoured recovery and recycling and restricted incineration and landfilling operations in the waste management. There are expectations that by banning landfilling, this can stimulate the closed loop of resources, which is also one of the CE argumentations (ZWIA 2015). As indicated by Cristina Ghinea and Maria Gavrilescu, in Chap. 3 of this volume, there are policy instruments that promote segregation and participation among the authorities, stakeholders and the local population, like a waste tax. It is mentioned that with a local tax, the funding for infrastructure, segregation schemes and communication campaigns will be able to be developed in order to generate a proper waste management system that would not follow a linear economy model.

It is then argued that Zero Landfill (landfill ban) is not necessarily considered as ZW. Even further, the ZWIA advocates that at least 90% of waste diversion from landfills, incinerators and the natural environment needs to be achieved. In addition to that, the commitment is to implement a strategy in order to transform the system into zero discharges. With this parameter of quantification and the 10 ZW business principles, the ZWIA is helping to guide and to evaluate the performance of policies and programmes executed by businesses. This added to the commitment to the triple bottom line (sustainability), following the precautionary principle. Buying reused, recycled and composted resources, responsible conduct and zero waste are some of the business principles that are embedded in the ZWIA vision (ZWIA 2015). All of these contribute to a combined effort to realise their ambition of diversifying at least 90% of waste from landfills, incinerators and the natural environment.

As zero waste is looking into the avoidance of unneeded use of materials and the optimisation of resources, there are several alternatives that are based in the following hierarchy: rethink, reduce, reuse, recycle/compost, recover, residuals management and leaving what it is unacceptable (ZWIA 2015). This hierarchy is also founded in the circular economy principles that are looking for the interaction (collaboration) among the different members from the value chain or stakeholders.

The collaboration between stakeholders is essential in order to apply a strategy of zero waste as circular economy. There are plenty of examples in the CE100 member list from the Ellen MacArthur Foundation about collaborations and partnerships between different members of the value chain (Ellen-MacArthur 2017). To illustrate, "Closed Loop[2]" is an example of the development of strong partnerships with their customers to perform waste management solutions with the best possible outcomes in terms of sustainability (Closed-Loop 2017). According to the Ellen MacArthur Foundation, Closed Loop is a company aligned with the zero waste and circular economy principles. In fact, "Closed Loop's approach is based on the principle of controlling material inputs to maximise recycling and recovery of materials, minimise waste to landfill whilst greatly reducing the environment footprint. An example of its activity would be the provision of recyclable packaging alongside a recycling programme, or the installation of an on-site food composting machine used to create fertiliser" (Ellen-Macarthur 2017). For running this project, there were diverse stakeholders involved who were coordinated by the project promoter, engaging among others local governments, public participation, recycling companies and technology developers.

There are different management and economic instruments that could help to support the structure of zero waste and zero waste to landfill policies, as is the case of the ones proposed by Zaman in 2015 or the "6 Rs". These were part of the proposal presented by the Recycling Council of British Columbia (RCBC). The RCBC is adopting a ZW vision in order to integrate the practices required from the parties involved in this type of policies (RCBC 2014). Even though there is a clear difference between ZW and ZWTL as is reported here, authors of some of the chapters of this volume use them indistinctively. What is important to stress is that their cases were framed under CE, which has more similarities with ZW.

[2] Company operating in Australia, New Zealand and the United Kingdom

1.3 Conclusion and Brief Introduction to the Content of This Book

This chapter lays out the conceptual foundations underpinning our selection of contributions and attempts to bring some epistemological clarity to the field, highlighting tensions between competing concepts.

We present papers that show the relevance of having one of the stakeholders[3] as the initiator of the implementation of initiatives under CE and/or ZW principles. Moreover, the engagement and consensus among different stakeholders' interests were explained in most of the cases. They need to be aligned to find ways to create collaborative schemes in order to maintain the resources' values as long as their quality and financial conditions permit it. Some of the cases here presented studied the CE and/or ZW theoretical frameworks trying to apply it/them to their local or regional context. These chapters analyse CE practices within their specific geographical contexts: Mexico, Indonesia, China, Romania and the Netherlands.

We use the type of stakeholder as the basis to structure this book and introduce the papers highlighting the role of the public sector as CE promoter (Chaps. 5, 6, 7 and 8) followed by initiatives coming mainly from the private sector (Chaps. 9, 10 and 11) and the civil society (Chap. 12). But before we fully enter to highlight the stakeholder role of those papers for this book, we present the first three papers (Chaps. 2, 3 and 4) that describe the challenges and opportunities that CE and ZW meet in their specific and/or broader context. This last aspect of context is clearly discussed in Chap. 4, in which the authors look through international frameworks such as climate change mitigation, in their search to identify potential opportunities.

References

Benyus J (1997) Biomimicry: innovation inspired by nature. William Morrow & Co, New York

Brezet H, Van Hemel C (1997) Ecodesign: a promising approach to sustainable production and consumption. UNEP, Paris

Chertow MR (2000) Industrial symbiosis: literature and taxonomy. Annu Rev Energy Environ 25:313–337

Chertow MR, Lombardi DR (2005) Quantifying economic and environmental benefits of co-located firms. Environ Sci Technol 39(17):6535–6541

Closed-Loop (2017) Closed loop – love what you do. http://www.closedloop.com.au/uk/about. Accessed 20 July 2017

Connett P, Sheehan B (2001) A citizen's Agenda for zero waste. http://archive.grrn.org/zerowaste/community/activist/citizens_agenda_2_read.pdf

Dresner S (2002) The principles of sustainability. Earthscan, London. Accessed 27 Aug 2017

[3] Stakeholder: any group or individual who can affect or is affected by a decision or action (Freeman 1984)

Ellen MacArthur Foundation (2013) Towards the circular economy volume 2: opportunities for the consumer goods sector. https://www.ellenmacarthurfoundation.org/publications/towards-the-circular-economy-vol-2-opportunities-for-the-consumer-goods-sector. Accessed 2 May 2017
Ellen MacArthur, Foundation (2017) Ellen MacArthur Foundation. CE100 Members. https://www.ellenmacarthurfoundation.org/ce100/directory/closed-loop. Accessed 27 Aug 2017
Freeman RE (1984) Strategic management: a stakeholder approach. Pitman, Boston
Frosch RA (1992) Industrial ecology: a philosophical introduction. Proc Natl Acad Sci USA 89:800–803
Geng Y, Doberstein B (2008) Developing circular economy in China: challenges and opportunities for achieving "leapfrog development". Int J Sust Dev World 15(3):231–239
Graedel TE (1996) On the concept of industrial ecology. Annu Rev Energy Environ 21:69–98
Haugh H (2005) A research agenda for social entrepreneurship. Soc Enterp J 1(1):1–12
Lansink A, de Vries H (2010) in 't Veld, De Kracht van de Kringloop; Geschiedenis en Toekomst van de Ladder van Lansink (The power of closing the loop; history and future of the ladder of Lansink). Vrilan, Apeldoorn
McDonough W, Braungart M (2002) Cradle to cradle; remaking the way we make things. North Point Press, New York
Murray A, Skene K, Haynes K (2015) The circular economy: an interdisciplinary exploration of the concept and application in a global context. J Bus Ethics 140(2):1–12
RCBC Zero Waste Hierarchy (2014) Recycling Council of British Columbia. https://www.rcbc.ca/files/u7/RCBC%20Zero%20Waste%20Hierarchy%20-%20APR%2014.pdf. Accessed 2 Aug 2017
Schneider A (2014) Reflexivity in sustainability accounting and management: transcending the economic focus of corporate sustainability. J Bus Ethics:1–12
Yuang Z, Shi L (2009) Improving enterprise competitive advantage with industrial symbiosis: case study of a smeltery in China. J Clean Prod 17(14):1295–1302
Zaman AU (2015) A comprehensive review of the development of zero waste management: lessons learned and guidelines. J Clean Prod 91(15 March):12–25
Zhang L, Yuan Z, Bi J, Zhang B, Liu B (2010) Eco-industrial parks: national pilot practices in China. J Clean Prod 18(5):504–509
ZWE (2015) Zero waste to landfill and/or landfill bans: false paths to a circular economy. Zero Waste Europe. http://www.zerowasteeurope.eu/2015/11/press-release-landfill-ban-a-false-path-to-a-circular-economy/. Accessed 31 Mar 2016
ZWIA (2015) Zero waste international alliance. http://zwia.org/aboutus/. Accessed 1 Sept 2017

Chapter 2
Potentials and Challenges for a Circular Economy in Mexico

Hans Dieleman and María-Concepción Martínez-Rodríguez

Abstract This chapter deals with the potentials and challenges – the strengths and weaknesses – of Mexico in moving towards a circular economy. In answering this question, the chapter does not look at reduction, reuse and recycling of waste in a narrow sense but approaches the topic from a wide and systemic perspective inspired by the concept of national innovation systems. The authors first provide some key data with respect to the current handling and management of waste in Mexico and then present in a conceptual way their systemic view on a circular economy. Subsequently they explore the potentials and challenges for Mexico, revising the following elements of their conceptual model: "market trends and conditions"; "competitiveness and productivity"; "the political and regulatory framework"; "education, training and knowledge transfer"; and "learning culture". They finish off with some conclusions concerning the potentials and challenges of Mexico in moving towards a circular economy.

Keywords Mexico · Circular economy · National innovation system · Learning culture · Legislation and politics · Education and training · Market conditions · Competitiveness

2.1 Introduction

What are the potentials and challenges – the strengths and weaknesses – of Mexico in moving towards a circular economy? We will try to answer this question, not by looking at reduction, reuse and recycling of waste in a narrow sense but by looking at a circular economy through a wide and systemic perspective (Leendertse 2016).

H. Dieleman (✉)
Urban Studies Center, Autonomous University of Mexico City, México City, Mexico
e-mail: johannes.dieleman@uacm.edu.mx

M.-C. Martínez-Rodríguez
Centro Interdisciplinario de Investigaciones y Estudios sobre Medio Ambiente y Desarrollo, Instituto Politécnico Nacional (IPN-CIIEMAD), Unidad Zacatenco, México City, Mexico

M.-L. Franco-García et al. (eds.), *Towards Zero Waste*, Greening of Industry Networks Studies 6, https://doi.org/10.1007/978-3-319-92931-6_2

We are convinced that a country's general economic conditions, cultural features, history, educational system and learning capacities strongly affect and determine its capacity to innovate, change and adapt to new conditions, opportunities and challenges. For Mexico, this is not different, yet the answer is necessarily multifarious, as the country is extremely multilayered and heterogeneous. Mexico is a complex country that has one foot in modernity and the other still in a pre-modernity. This strongly affects the country's possibilities to transform into a circular economy. To give an idea of what this means, we start off with a rather long citation of Philip et al. (2015) (with some modifications), as they made a good characterisation of Mexico:

> On the one hand, Mexico's elites – business, governmental and academic – are all generally advanced in their outlook. Many of Mexico's policymakers are at home in Western institutions, and numerous Mexicans now have degrees from American, Canadian and European universities. In addition, at least a portion of the tertiary education system is well advanced in terms of international competitiveness. On the other hand, Mexico suffers from many problems that are not typical for OECD countries. Challenges include a wasteful and over-unionised education system, extreme poverty, rampant corruption, rampant organised crime, chronic tax evasion and clientelistic governance at the state and municipal levels. [Philip et al. coordinators (2015:2)]

There is a well-educated – elitist – minority in Mexico that wants to change towards a modern sustainable country, but there is also a large majority that is either apathetic or overtly resistant to change. This resistance is fed by the emotion of missing the necessary capacities to work under conditions of modernity and results in a rejection of modernity rather than in the desire to acquire the skills needed to function under conditions of modernity. This strongly affects the country's learning capacity, flexibility, productivity and economic competitiveness. It is this image of a double face that will emerge in the following paragraphs, when assessing the strengths and weaknesses of Mexico in moving towards a circular economy.

2.2 A First Glance at Mexico's Waste Management Situation

In 2012, Mexico produced 103,000 tons of wastes per day – 905 g per day per Mexican – making the country the 10th largest producer of solid waste in the world (SEMARNAT 2012). Data provided by INEGI, the National Institute of Statistics and Geography, show that the total production of municipal solid waste has increased by 33% in the period between 2001 and 2013 (Góngora Pérez 2012). This constitutes an increase of 2.7% annually in that period. The country actually has 650 open-air dumpsites and about 200 landfills, and many of them are in poor conditions and lack basic infrastructure to ensure a sound operation and monitoring of the waste streams (Gasnier and Portales 2008). In 2008, it was estimated that 67% of the total municipal solid waste generated in Mexico was disposed of in sanitary landfills and controlled sites, while the remaining 33% was disposed of in uncontrolled ways (Gasnier and Portales 2008). In 2009, the government developed financial support programmes designed to help modernise the country's waste management system. Grants were issued to upgrade to landfills, close open sky dumps and engage in technical landfill studies, municipal waste collection and the development of state, municipal or inter-municipal waste management plans. Many governmental agencies, however (mainly local governments), continue to see waste management as a low priority, and many industries, especially the smaller ones, are hardly stimulated to reduce, reuse or recycle (Velasco Pérez Alonso 2011).

According to data of SEDESOL, the Mexican Ministry of Social Development, the economic potential for recycling of waste in Mexico is over $1.34 million USD daily, while data provided by INEGI, the National Institute of Statistics and Geography, shows that the existing recycling businesses generated $630,119 USD per day, which represents just 47% of its full potential (ITA 2016). Despite of this potential for recycling, the practice of reduction, reuse and recycling of wastes is developing slowly.

Mexico is far away from having a circular economy, and it will take time and effort in many sectors of the Mexican society to transform into a circular economy (Perry-Gottesfeld and Durand 2011). Such a transition process will not only require changes in recycling and reuse of waste; it equally requires changes in production and consumption, and in the institutional systems surrounding production and consumption, such as in legislation, education, knowledge transfer, environmental awareness, entrepreneurial cultures and more. In this chapter, we explore the possibilities of a transformation process towards a circular economy in Mexico, through such a systemic lens inspired by the concept of national innovation systems (Lundvall 2010).

2.3 A Circular Economy Seen as Entirety of Production and Consumption with a Surrounding Support System

A circular economy can best be seen as a complex system with three basic parts: production, consumption and the surrounding support system (RSA Innovate 2016). Each of these parts has its own elements and characteristics, and the interaction of them determines the chances for Mexico to transform into a circular economy. That first part encompasses the core concept of a circular economy. By this, we refer to the actual changes in industrial production through rethinking how linear manufacturing models can become circular closed-loop models (Yuan et al. 2008; Ellen MacArthur Foundation 2013). It implies seeing material flows in an economy as being part of cycles and more in particularly of two basic cycles with distinct characteristics: biocycles and techno-cycles. With respect to the biocycle, the objective is to make biomass return into the biosphere after product use, in direct ways or in a cascade of consecutive use. With respect to the techno-cycle, which is built up of inorganic products and materials such as metals and plastics, the strategy is to keep them in closed loops to ensure the possibility of reuse and recycling and to prevent potential pollution (Jackson et al. 2014; Cramer 2014).

One of the first to conceptualise the idea of a circular economy was the American economist Kenneth E. Boulding (1966) in his article, "The economics of the coming spaceship earth". We live on a spaceship, he argued, with limited natural resources, and because of that, we cannot maintain our linear "take-make-dispose" economy. We must find ways to bend the linear chain of production, distribution and consumption in a circular one. Since then, many concepts were introduced and tested that all touch upon parts of a circular economy, such as the concepts of cleaner production, industrial ecology, eco-design and design for reassembling and socially responsible entrepreneurship. In 2002, Michael Braungart and William McDonough made an important contribution to the circular economy in their book "Cradle to Cradle: Remaking the Way We Make Things", discussing the idea of an economy that works in loops with positive impacts on economic competitiveness, job creation, resource savings and waste prevention (McDonough and Braungart 2002). All these concepts question linear manufacturing models and are part of the core of the concept of a circular economy.

The second part of the system "circular economy" deals with consumption and consumption patterns. The main driver behind industrial production is consumption with particular market dynamics as cultural needs, marketing influences and market trends. These dynamics are important positive or negative factors in a transformation process towards a circular economy. Today, marketing is largely a negative factor as it predominately stimulates a culture of consumption telling us that we are what we consume and what we possess (Raworth 2012). Ecofriendly marketing exists as well but remains in the margin of mainstream marketing efforts. The globalisation of industrial markets offers products of low prices distributed worldwide and often stimulates a fast rate of consumption of non-durable consumption goods. It can however also have potential positive impacts on a circular economy, as we will show in the case of Mexico, later on. Generally speaking, many trends in consumption go against a circular economy, so strategies and policies are needed to curve this trend.

Three different strategies can be identified to make consumption fit into a circular economy: (1) addressing products, (2) persons and (3) functions and markets (Persson 2015; Boulanger 2016). The strategy to address products fits into that what Boulanger calls the *eco-efficiency* perspective, and aims at diminishing the environmental impacts of the use and possession of products, mainly through redesign and creating systems for reuse and recycling. The second strategy fits into what Boulanger calls the *sufficiency* perspective touching upon the person instead of the product. The logic of sufficiency consists of consuming the right quantity of material goods and services, which escapes "both the Charybdis of under consumption (poverty) and the Scylla of overconsumption" (Boulanger 2016:5). This strategy is mainly cultural and touches upon attitudes and orientations of people. It is realised through education, information and enhancing awareness. The third strategy aims at the creation of a *sharing economy* where the conventional ownership of things is being replaced by schemes of "bartering, lending or renting" (Botsman and Rogers 2010, p. xv). Here, the function of products becomes central, rather than the ownership of products. This strategy is conceptual as it aims at reformulating both what a "product" is and what a "market" is. It aims at changing the interaction between consumers, retailers and manufactures around new concepts such as "performance-for-pay models", "rent or leasing schemes" and "return and reuse practices" (EMF 2013, p.10). These three consumption-oriented strategies are crucial in a transformation process towards a circular economy.

The third part of the system "circular economy" is the surrounding support system of production and consumption and is made up by various cultural and structural characteristics of a country or region. Stimulating innovation and transformation towards a circular economy requires a certain political and regulatory framework that gives incentives, set conditions and interferes, so actors change behaviour facilitating the innovation and transformation looked for. Regulations create conditions, and policies may indicate directions, which are all needed to create an environment of change into a certain direction. Economic actors equally need to be educated and trained and because of that, a properly functioning knowledge transfer system is a crucial part of the surrounding support system of a circular economy.

Cultural characteristics of an economy are crucial in understanding the way the economy moves, functions and performs (Tabellini 2010) and are equally crucial in determining the potential for change and transformation into the direction of, in our case, a circular economy (Lundvall 2007). Such cultural features can be found both on a macro and a micro level and are usually mixed, manifesting themselves on various levels simultaneously. Entrepreneurial cultures in companies, for instance, differ in some ways but are at the same time largely influenced by national cultures. Finally, the essence of support systems is not found in the mere existence of institutions, regulations or structures but, as the OECD rightfully argued, is to be found in the linkage and interaction among actors. It is the complex or set of relationships among actors that produce, distribute, consume, generate knowledge, educate and apply knowledge in various concrete practices, which create the strength of the support system (OECD 1997). This is the cultural complex of capacities, attitudes and orientations, which determine the strengths of a country to innovate and transform.

2.4 Mexican Potentials and Challenges for a Circular Economy

What are the potentials and challenges – the strengths and weaknesses – of Mexico in moving towards a circular economy? In answering this question, we explored various elements of the here above presented complex set out. Those elements were put together in Fig. 2.1. Based on these explorations, we made a qualitative assessment of the potentials and challenges for Mexico to move towards a circular economy. The assessment is presented in Sect. 2.2 of this chapter.

2.4.1 Market Trends and Conditions

The Mexican economy has been relatively stable over the past decades, with an average growth of 2.6% in the past 10 years (Pèrez 2015). Based on this growth pattern, and compared with growth patterns in other large countries, Mexico is projected to become the world's 8th largest economy in 2050, before Russia, the UK and France (Pèrez 2015). Mexico's geographical location favours economic

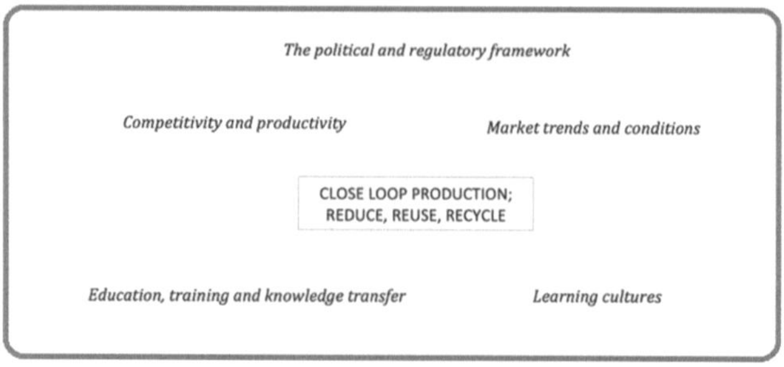

Fig. 2.1 Relevant elements in the transformation to a circular economy (contribution of the authors)

development, as it is rather centrally located in between most of the world's economic centres. The average flying time from Mexico City to New York, L.A., Rotterdam, Yokohama, Río de Janeiro, Shenzhen and Shanghai is 14.4 h, while the average flying time for European and Asian cities to fly to the other major economic centres is between 20 and 25 h (Pèrez 2015). The dependency on the American economy however makes the Mexican economy vulnerable to changes in political relationships between the two countries, as well as to fluctuations in the American economy (Moreno-Brid et al. 2005). A diversification of the economic relationships of Mexico with other parts of the world is therefore needed to ensure a safe economic growth in the coming decades. The overall economic situation of Mexico, and the trends for the future, does not show obvious hurdles for a transformation towards a circular economy.

Mexico still is predominantly a country of manufacturers, complemented with tourism, mining, agriculture, real estate and construction, while the financial and service sectors are gradually gaining importance in the national economy (PwC 2014). Industrial production and manufacturing is realised in the rather famous Mexican *maquiladoras*, the assembly industries. These maquiladoras are concentrated in the northern part of the country, in the relative proximity of the US border, but they are obviously also found in other parts of the country, though in lesser concentration. Tourism has an important concentration in the Yucatan region and especially the Cancun area. But again it must be added that tourism is found all over Mexico, but with less regional concentration. The upcoming financial and service sector is largely found in the big cities, with a significant concentration in Mexico City.

An important characteristic of manufacturing, tourism and finance is their international character, with perpetual flows of import and export and travel of people in and out of Mexico. We see this as a potential for a transformation process towards a circular economy. The Mexican industrial sector, for instance, is manufacturing, as a single nation, a significant part of the world's cars (trucks 5.4% and light vehicles 8.1%), TVs (4.2%), cell phones and communication equipment (3.9%), car parts and accessories (5.7%) and computer parts (5.2%) (Pèrez 2015).

Design and product specifications of cars, computers and TVs, but also of food products such as beer, ice cream or chocolate, take place outside of Mexico in the headquarters of companies like Chevrolet, Volkswagen, Nissan, Dell, Nestlé, Unilever, Danone or Heineken. All these companies are present in Mexico and prescribe to Mexican manufacturers how to produce, manufacture and assembly according to their product and production specifications. Mexico is in this respect a follower and will without any doubt follow when companies decide to change the design of their products and production processes to fit circular economy pattern.

The tourist sector, especially in the Cancun area, has similar characteristics, as foreign hotel chains dominate the sector, with names as InterContinental, Marriott, Ritz Carlton, Fiesta Inn, Ibis or Hilton. In many of these chains, environmental management standards are company-wide established and internationally maintained, while changes in standards and requirements are made in the headquarters of the companies. Here as well, we see potential for a change towards circular economy, even if there is little incentive to change towards a circular economy inside of Mexico. The financial and service sector strongly represents the modern world and introduces in Mexico those values and lifestyles that represent the modern world. This has both positive and negative impacts. In Mexico City, representatives of this sector were the first to embrace the bike-sharing system introduced some years ago and equally are promoters of car-sharing systems and the consumption of organic food. On the other hand, they frequently fly which results in ecological footprints that are certainly not sustainable. In general, we favour the cultural impact they have on their country.

Mexico is not only just following however. There is a growing interest in applying environmental technology in all sectors of the Mexican industry, mainly due to more stringent environmental legislation. This creates a demand that is largely satisfied through importing technology from other countries, mainly Europe and the USA, with the USA holding the largest stake in Mexico's import of environmental technology. Export of environmental technology from the USA to Mexico has been growing rapidly in the past years. The US International Trade Administration talks about "an unprecedented investment in environmental infrastructure, that gave Mexico an upward mobility by improving two points in the world ranking in the Environmental Technologies Top Market Study" (ITA 2016: 20). Mexico gained this position, immediately following China, thanks to investments in air pollution and the water sector, and to a much lesser degree, thanks to investments in waste management. Technologies with a high level of application in air pollution are, among others, continuous emission monitoring systems, selective catalytic reduction systems, electrostatic precipitators and catalytic converters. In the water sector, a whole range of technologies is being imported nation-wide, as, for instance, water reclamation technologies, water reuse equipment and services and advanced filtration technology such as membrane filtration, reverse osmosis, UV disinfection and anaerobic digestion.

In the sector of waste management, however, Mexico is still not doing well. While the country holds the number 2 position overall in the import of environmental technology from the USA, it is ranked only number 13 in imports of waste management technology. Waste management technology imported from the USA includes waste handling equipment, biogas capture technologies, sanitary landfill systems and sorting machines.

2.4.2 Competitiveness and Productivity

While the market conditions are generally speaking rather favourable for Mexico, the country's competitiveness and productivity are not. This generates a negative incentive for a change towards a circular economy, just as the positive market conditions create a potentially positive impact (Sánchez 2014). A country's competiveness is usually seen as its capacity to produce high quantities of product, in time and with high quality, and to be flexible, creative and innovative with good capacities to enter new markets and adapt to new ways or conditions of production. Competiveness depends on both macro and micro factors (Békés-Gábor 2015). Macro factors are the availability of infrastructure (roads and IT infrastructure) and of capital (accessibility to credit), the flexibility of the market and the quality of regulation and of education and research. We will address regulation and education in subsequent paragraphs and will therefore waive these issues here. Micro factors are the presence of an entrepreneurial culture, a skilled workforce and participation rates in post-school education (Martin 2004).

The availability of IT infrastructure has been a challenge for Mexico for a rather long time, but considerable improvements have been observed in the last decade (Hanson 2011). The poorly functioning credit market however remains a critical factor in Mexico's competitiveness. Over the past decade, Mexico managed to open the market for private credits to small firms and households but still struggles with issuing long-term credits for investments in large-scale infrastructural projects (Hanson 2011). The Bank of Mexico concludes that Mexico's competitiveness is very low, due to the country's market rigidity, too much regulation and a too strong influence of unions (Chiquiar and Ramos-Francia 2009). The bank promotes political reform that typically fits into a neo-liberal discourse yet touches in Mexico on number issues that are widely perceived as real problems. This is first of all the overregulation of the Mexican market, an abundance of regulation that is supposed to be enforced by often incompetent and corrupt governments. As a response to this, many companies choose to operate within the informal economy, and the informality of the Mexican economy is widely seen as a major factor in the country's lack of competitiveness (Hanson 2011; Chiquiar and Ramos-Francia 2009). A second factor is the strong position the unions have in both the private and the public sector. Though unions in principle play an important and positive role, Mexican unions tend to block any form of change, not because of the proposals as such but because of the type of resistance mentioned in the introduction of this chapter. This resistance is rather cultural than economic of operational.

On the micro level, a key component to competiveness and productivity is a culture that stimulates initiatives and embraces new ideas. Such a culture, however, is weakly developed in Mexico. Gordon (2010) analysed Mexican business cultures using the dimensions that Hofstede developed in his famous cross-cultural analysis of 50 countries (Hofstede 1980). These dimensions are individualism versus collectivism, power distance, uncertainty avoidance and masculinity versus femininity. Almost all of these dimensions – with the exception of masculinity versus femininity – explain why Mexican entrepreneurial cultures do not favourite competiveness and productivity. The idea of making career through hard work and showing that you are more competent than your co-workers clashes with the collective culture of Mexico. Mexico has a "being" rather than a "doing" culture, where relationships are more important than work and where "unfinished tasks or jobs are not a worrisome thing" (Gordon 2010:9). Mexicans equally have, on average, a low tendency of uncertainty avoidance, meaning that they accept life as it comes, rather than working hard in an attempt to control and shape future developments. The phrase "What will be, will be" or in Spanish "*Que sera, sera*" is rather exemplary for this attitude or culture (Davis and Nayebpour 2004).

Traditional Mexican management styles are based on punishment rather than on stimulation or empowerment. As a result, Mexican workers try to avoid taking responsibility and are often more comfortable with a low profile than a high profile. While doing an anthropological study in one company, Gordon discovered that the word "empowerment" did not exist in the vocabulary of the managers and that they had difficulties in understanding the concept, as in translating it into Spanish. This all is not helping the creation of entrepreneurial cultures where creativity, initiative and new ideas flourish and are easily embraced.

Culture matters, as Guido Tabellini demonstrated through his research in different economic regions in Europe. His research findings indicate that culture and a particular cultural history strongly correlate with regional economic development and explain quite well why some regions develop faster than others (Tabellini 2010). Because of this, we look at Mexican business cultures as a real challenge for a transformation process towards a circular economy. (See also Lewandowski 2016).

2.4.3 The Political and Regulatory Framework

Mexico has 32 states, Mexico City recently being included, and has about 2454 municipalities. Its legal system is based on the European, and especially French, system, with a division between the legislative, executive and judicial branch. The legislative branch is comprised of a congress with a chamber of deputies and chamber of senators. The executive branch – the president, ministers and their administrative entities – is responsible for executing the laws enacted by the congress, and the judicial branch has the function of interpreting the laws and resolving disputes. On the federal level, the key administrative body responsible for environmental affairs is the Secretariat of the Environment and Natural Resources (*Secretaría de Medio Ambiente y Recursos Naturales*, SEMARNAT). Many environmental areas have special agencies with far-reaching authority to develop and execute politics, such as the National Water Commission, the National Commission for Protected Natural Areas or the National Institute of Ecology and Climate Change. Other areas are decentralised, such as municipal waste management that largely is the responsibility of local and municipal governments. The differentiation and decentralisation of politics are seen as a positive development, but what is missing is a proper coordination between different institutions and levels of government (Gasnier and Portales 2008).

Mexican politics, in general, has been going through a process of change since the end of the previous century. Policy development was heavily influenced by corporatism, meaning that politics was developed by a conglomerate of big industries and politicians from the PRI, the Institutional Revolutionary Party, which ruled the country since the Mexican Revolution until the year 2000 (Diez 2006). The liberation of the political system in the year 2000 seemed to make an end the old form of corporatism and promised to democratise the policymaking process to some extent. It is hard to predict what the future will bring however, as tendencies to return to the old system are clearly present (Loxton 2016).

Another key characteristic of Mexican politics and regulation is the existence of the huge gap between policy formulation on the one hand and implementation and enforcement of regulation on the other hand (Diez 2006). In the environmental field, this accounts for a serious lack of enforcement of legislation, with many companies operating without the proper permits to do so. Clientelism and corruption deepen this problem, as does the informal character of the Mexican economy. In practice, the gap between norm and deed is a real hurdle for the Mexican society to change towards environmentally more friendly ways of production, distribution and consumption. This is especially true for medium and small companies. The larger companies, and certainly the multinational ones, conform rather frequently to their own internationally applied standards, irrespective of the political/regulatory framework inside of Mexico.

Despite of the obvious flaws in the Mexican political/legislative framework, there are also positive developments to be mentioned. Since some years, Mexico is developing a National Development Plan (NDP) to drive developments in a certain direction with clear objectives and goals. The National Development Plan 2013–2018 was launched by president Enrique Peña Nieto, who´s presidency worked to make profound changes to the country in five distinct areas:

1. To achieve peace in Mexico that will advance democracy and security
2. To make Mexico more inclusive to all citizens, while protecting their social rights
3. To improve the quality of the education system so that youth in Mexico can meet the challenge of an ever-more competitive world
4. To promote prosperity by stimulating economic growth in a way that Mexicans will feel the prosperity directly in their pockets
5. To consolidate Mexico as a responsible international player, defending international law and promoting free international trade, showing its solidarity with other countries

The words environment, sustainability or circular economy are not found among the five key objectives, but they are present as objective 4.4, under the heading of prosperity. Objective 4.4 of the national plan aims at stimulating "green growth", while the way to operationalise this is published in the "Sectorial programme for environment and natural recourses 2013–2018" (PROMARNAT), developed in parallel with the overall national plan.

The PROMARNAT plan presents a comprehensive diagnose of the current environmental situation in Mexico, in the areas of air, water, waste and biodiversity, followed by various concrete plans and objectives. For waste, the objective is to reduce the percentage of uncontrolled disposal of waste from 30% in 2013 to 17% in 2018. This must be realised through various policy plans, such as plans to improve integral solid waste management on the municipal level and to create more infrastructures needed to manage waste in municipal dumpsites. This reduction of the percentage of uncontrolled disposal of waste is a serious step forward towards a circular economy, even though lots of subsequent steps are obviously needed.

A positive development as well is the engagement of the private sector, the international character of the Mexican economy as mentioned before and the support given by other countries in this respect. One interesting example is the seminar "Emerging Mexican Circular Economy Market, Possibilities for Finnish Companies", organised on March 7, 2017 by the Finnish-Latin American Trade Association. The seminar explored concrete option of collaboration around topics as biogas projects at landfills and waste-to-energy projects within energy-intensive industries, such as cement, steel and petrochemical industries.

2.4.4 *Education, Training and Knowledge Transfer*

Seventy percent of all children in Mexico attend a public school, while 30% attend a private school (Barraza 2001). Due to the vast population growth, total student enrolments in public schools increased from 3.25 million students in 1950 to 28.22 million students in 2000, up to 34.8 million students in 2011 (Clark and Monroy 2013). As budgets for public education increased only at a margin of the increase in students, basically all public schools face budget problems and an abundance of students in their classrooms. The average number of students in one classroom is about 35 (in Mexico City frequently up to 50), and this obviously affects the quality of education (INEE 2005). A second characteristic of Mexico's public educational system is its traditional pedagogical orientation, emphasising mere knowledge transfer instead of teaching through active involvement of the students in their learning process. Thirdly, the average competence of teachers is low, because many are not selected and contracted based on their knowledge and teaching skills but because they are friends or family of the union representatives in charge of the selection process. The 2013 educational reform aims at improving the quality of the curricula, the teaching and teachers, and seeks to improve selection and wages of the teaching personal (Levinson 2014). The outcome of this reform may be positive, but this depends really on the way it will be executed and the budgets that will be allocated to ensure a sound execution.

Since the 1980s, environmental education is part of the curricula in all schools in Mexico, and each of the 32 states has a state-wide plan for education, training and communication in environmental matters (Marcos-Iga and William 2011). Many cities today have centres specialised in environmental education, and basically all universities offer a bachelor or master's programme in one or more environmental specialisations, like environmental education, environmental engineering, environmental planning or environmental law (Cecadesu 2006). There are however some serious restrictions, mainly due to a lack of innovation in teaching methods and teaching orientations (Ruiz et al. 2009). Most environmental education focuses on isolated environmental issues as water, waste or air pollution and lacks a more systemic orientation of sustainability, linking the environment with economic and social issues (Cecadesu 2006). A holistic orientation, linking individual action with societal effects and consequences, equally is missing in most of the curricula (Paredes 2005). Paredes Chi observes that, like education in general in Mexico, environmental education has too much focused on mere knowledge transfer, and she suggests that students should become actively and more critically engaged in problem definition and identification of possible environmental actions to address these problems (Paredes 2005).

Many universities offer post-academic continuous education in environmental matters, targeted at adults and professionals. Such *diplomats* usually take 6 months of classes (up to 6 h per week) and are finished with the writing of an essay or thesis. The *diplomats* have academic value and are recognised by both the public and the private sector. They open opportunities for adults – who need to have as minimum a bachelor degree – to deepen or change their career and to specialise in a certain environmental field. Here once more however, the teaching method and orientation create some limitations. Many Mexican universities teach with a rather exclusive theoretical orientation and have little experience in trace disciplinary collaboration with nonacademic actors of the public or private sector. This seriously limits the potential of the *diplomats* in preparing the participants to apply what they learned in practical contexts.

Finally, Mexico has multiple platforms and networks for knowledge transfer in environmental matters and in sustainability, mainly initiated and maintained by the private sector. Mexico equally participates in various international networks, initiated by the World Bank, the OECD or the UN. An example of the last is the Green Growth Knowledge Platform, which provides general information on best practices both with an international and a national perspective. Examples of information disseminated are the results of a comparative study of green growth in Mexico, Brazil and Chile and information on urban-industrial environmental management in Mexico. The World Bank has a rather similar website providing both international and Mexican information on business and sustainability. Various national networks exist, as for instance, the "Programme for Green Jobs[1]", the "Mexican Association of Environmental Enterprises" or the "Green Network". These networks exchange experiences of Mexican companies in green entrepreneurship and provide information on Mexican legislation and special sustainable projects in the country.

[1] http://empleosverdes.mex.ilo.org/empresas-empleadores/que-son-los-empleos-verdes

2.4.5 *Learning Culture*

The final issue to address is that of Mexico's learning culture, the country's capacity of adapting to new practices or technologies. On one hand, Mexicans are very quick in adapting to new technologies and new technology-induced practices. In this context, the experience of an American educator who visited Mexico City is rather illustrative. I was surprised to find the following in the streets of Mexico City, he said: "the taxi driver relied on Google Maps to navigate the chaotic network of streets, and driving from the airport. Public Wi-Fi spots are commonplace and their number is growing. When I checked for a Wi-Fi connection at traffic lights there would usually be many, many domestic Wi-Fi spots" (Grove 2014). Mexico City, as many other cities in the country, shows a face of fully being adapted to the modern world. Mexico has a collective culture where being connected to others is a condition "*sine qua non*" for life in general and survival in particular. Today, most Mexicans are constantly connected through Facebook, WhatsApp, Instagram and other social media.

In other ways though, Mexicans are not so open for change. Most Mexicans stick to their traditional extended family life and traditional ways of spending leisure time, and these two largely overlap: leisure time is spending time within the extended family. In professional setting, many refuse to adapt to modern forms of communication and collaboration. They apply the family model in professional settings and have problems distinguishing between the personal and the professional (Gordon 2010). Critique expressed in professional settings is not perceived as functional input to improve performance but is rather soon perceived as personal attack. What is missing is a culture of reflectivity where people, either individually or in collectively, critically interrogates their actions, with the intention to learn from experience and improve professional performance. This seriously hampers collaboration in – especially horizontal – teams and networks and poses a real challenge for a transformation process towards a circular economy, as the support system for such a transformation depends on the interaction, collaboration and reflective actions of the institutions and actors within the system (Jakonis 2010).

2.5 Conclusion

As mentioned in the introduction, Mexico is a complex country with at least two rather different and often contradicting faces: a modern and a premodern one. These two together define Mexico's chances to transform towards a circular economy. The market conditions and long-term market trends in the country are positive and should be seen as a potential positive condition for a – long-term – transformation process towards a circular economy. The country's National Development Plan aims at stimulating green growth and recognises the need to improve the existing infrastructure for reuse and recycling of waste. The Mexican Ministry of Social

Development identified a huge economic potential for recycling of waste, and governmental funds are liberated to upgrade state, municipal or inter-municipal waste management plans. Environmental education exists since the 1980s, and each of the 32 states has a state-wide plan for education, training and communication in environmental matters. Knowledge transfer systems equally exist, on both national and international levels, and, in general, Mexicans are keen to adapt themselves to new technologies and new technology-induced practices. Most of these indicators provide a positive outlook on Mexico's potential to change to a circular economy and show that many opportunities exist.

The question is if Mexico will be able to seize these opportunities. This question is real, as a number of serious challenges exist as well. Markets are rather inflexible; banks refuse to provide long-term credit; and above all, many organisations, both in the private and the public sector, lack practice that favours new ideas, innovation and creativity. A collectivistic culture of "being" rather than "doing", combined with a low perceived need to reduce uncertainty and a related desire to shape and control the future, leaves Mexico with a mentality of "*Que sera, sera*" or "what will be, will be". Such a mentality does not create incentives to seize opportunities. The discrepancy between the creation and execution of policy plan is another real challenge. Good intentions and plans exist, but the Mexican history shows that realisation of plans is not easy. With respect to environmental education, something similar can be observed. Even though environmental education exists, its content is not really geared at learning-by-doing and reflective practice, and its potential in a transformation process towards a circular economy is therefore limited. Finally, Mexico really misses a learning culture that stimulates reflective practice and collaborations or co-creation in teams. This flaw in learning capability is a characteristic that permeates many spheres of economy and society and is one of the biggest challenges to overcome.

References

Barraza L (2001) Environmental education in Mexican schools: the primary level. J Environ Educ 32(3):31–36. https://doi.org/10.1080/00958960109599143

Békés-Gábor (2015) Measuring regional competitiveness: a survey of approaches, measurement and data. Műhelytanulmányok discussion papers. Institute of Economics, Centre for Economic and Regional Studies, Hungarian Academy of Sciences Budapest, 2015. Mt-dp – 2015/29. ISBN:978-615-5447-91-4, ISSN:1785 377X

Botsman R, Rogers R (2010) What's mine is yours: the rise of collaborative consumption. Harper Collins, New York

Boulanger PM (2016) Three strategies for sustainable consumption, S.A.P.I.EN.S. http://sapiens.revues.org/1022. Accessed 10 Apr 2017

Boulding K (1966) The economics of knowledge and the knowledge of economics. Am Econ Rev 16:1–13

CECADESU, Centro de Educación y Capacitación para el Desarrollo Sustentable (2006) Estrategia de educación ambiental para la sustentabilidad en México. SEMARNAT. http://cecadesu.semarnat.gob.mx/. Accessed 9 Apr 2017

Chiquiar D, Ramos-Francia M (2009) Competitiveness and growth of the Mexican economy. Banco de México. Working papers. No. 2009-11. http://citeseerx.ist.psu.edu/viewdoc/download?doi=10.1.1.622.7227&rep=rep1&type=pdf. Accessed 6 Apr 2017

Clark-Nick, Monroy C (2013) Education in Mexico. World Education News & Reviews, WES Group Manager for Latin America. http://wenr.wes.org/2013/05/wenr-may-2013-an-overview-of-education-in-mexico. Accessed 6 Apr 2017

Cramer J (2014) Moving towards a circular economy in the Netherlands: challenges and directions. Utrecht Sustainability Institute, Utrecht University and Economic Board Amsterdam, The Netherlands. http://usi-urban.wp.hum.uu.nl/files/2015/04/Paper-HongKong-JC-april-2014.pdf. Accessed 5 Apr 2017

Davis AS, Nayebpour MR (2004) Obreros (workers) against Gerentes (managers): changing values in the Mexican workplace. Lat Am Bus Rev 5(1):71–93

Diario Oficial de la Federación 12-Diciembre-2013. México

Díez J (2006) Political change and environmental policymaking in Mexico. Routledge, New York

Doing Business in Mexico, Turismo (2014) Guía para la inversión y desarrollo del turismo en México, PwC México

"Emerging Mexican Circular Economy Market, Possibilities for Finnish Companies" organised by the Finnish-Latin American Trade Association. http://www.inee.edu.mx/bie/mapa_indica/2005/PanoramaEducativoDeMexico/PG/PG01/2005_PG01__pdf. Accessed 10 Apr 2017

EMF Ellen MacArthur Foundation (2013) Towards the circular economy: opportunities for the consumer goods sector. http://www.mckinsey.com/~/media/mckinsey/dotcom/client_service/sustainability/pdfs/towards_the_circular_economy_emf_report.ashx. Accessed 10 Apr 2017

Gasnier A, Portales LE (2008) PETSTAR: Adding value to the chain of recycling and improving the scavengers' working conditions. Latin America & Caribbean, Mexico. Sector Waste. United Nations Development Programme. http://growinginclusivemarkets.org/media/cases/Mexico_Petstar_2008.pdf. Accessed 8 Apr 2017

Góngora JP (2012) El reciclaje en México. Comer Exter 64(3)

Gordon G (2010) Managing in Mexico – an ethnographic comparison to theory and previous research. J Int Bus Cult Stud 2

Grove J (2014) Researching mobile learning culture in Mexico city's schools. Resource document. McGraw-Hill Education. http://blog.mheonline.com/researching-mobile-culture-in-mexico-citys-schools. Accessed 10 Apr 2017

Hanson GH (2011) Understanding Mexico's economic underperformance. Migration Policy Institute, Washington, DC

Hofstede G (1980) Consequences: international differences in work-related values. Sage, Beverly Hills

INEE, National Institute for the Assessment of Education in Mexico (2005) Tamaño promedio de los grupos en primaria y secundaria. http://www.inee.edu.mx/bie/mapa_indica/2005/PanoramaEducativoDeMexico/PG/PG01/2005_PG01__pdf. Accessed 10 Apr 2017

Jackson M, Lederwasch A, Giurco D (2014) Transitions in theory and practise: managing metals in the circular economy. Resources 3:516–543

Jakonis A (2010) Organisational culture in multicultural organisations – Mexico. J Intercult Manag 2(2): 83–96. ISSN:2080-0150. http://jml2012.indexcopernicus.com/fulltxt.php?ICID=935412. Accessed 10 Apr 2017

Leendertse P (2016) The transition from a linear to circular economy: an innovation system analysis of the composites industry. Master's Thesis Internship-Sustainable Business & Innovation. Utrecht University, Faculty of Geosciences

Levinson BA (2014) Education reform Sparks teacher protest in Mexico. Sage Journals, Phi Delta Kappan. http://journals.sagepub.com/doi/pdf/10.1177/003172171409500811. Accessed 8 Apr 2017

Lewandowski M (2016) Designing the business models for circular economy—towards the conceptual framework. Sustainability 8(1):43–48

Loxton J (2016) Authoritarian successor parties worldwide: a framework for analysis. University of Sydney

Lundvall B-Å (2007) National innovation systems-analytical concept and development tool. Aalborg University and Tsinghua University. Conference in Copenhagen, Denmark

Lundvall B-Å (2010) National systems of innovation: toward a theory of innovation and interactive learning. Anthem Press, London/New York/Delhi

Marcos-Iga J, William S (2011) Current state of environmental education in Mexico: a study on practises, audiences, settings, and topics. Appl Environ Educ Commun 10(4):219–227. https://doi.org/10.1080/1533015X.2011.669687
Martin LR (2004) A study on the factors of regional competitiveness. A draft final report for the European Commission Directorate-General Regional Policy Cambridge Econometrics. University of Cambridge, Cambridge
McDonough, Braungart (2002) Cradle to cradle: remaking the way we make things. North Point Press, New York, p 208
Moreno-Brid JC, Santamaría J, Rivas-Valdivia JC (2005) Industrialization and economic growth in Mexico after NAFTA: the road travelled. Dev Chang 36:1095–1119. https://doi.org/10.1111/j.0012-155X.2005.00451.x
National Innovation Systems (1997) Organisation for economic co-operation and development. https://www.oecd.org/science/inno/2101733.pdf. Accessed 9 Apr 2017
Occupational Knowledge International. http://www.global-chemicals-waste-platform.net/fileadmin/files/doc/Case_Studies/Case_Study__Mexico_ULAB.pdf. Accessed 9 Apr 2017
Paredes CAA (2005) Theory and practise in environmental education: a Mexican case study. Thesis Doctor of Philosophy. Deakin University, Australia
Pèrez CF (2015) Circular economy & business opportunities in Mexico. ProMéxico
Perry-Gottesfeld, Durand K (2011) Case study Mexico: lead battery recycling
Persson O (2015) What is circular economy? – the discourse of circular economy in the Swedish public sector. The Department of Earth Sciences, Uppsala University Published at Department of Earth Sciences, Uppsala University (www.geo.uu.se), Uppsala, 2015. http://www.diva-portal.org/smash/get/diva2:841910/FULLTEXT01.pdf. Accessed 9 Apr 2017
Philip G, Faust J, Thunert M (2015) Sustainable governance indicators, Mexico report. Bertelsmann Stiftung
Plan Nacional de Desarrollo 2013–2018. Gobierno de la República, México. http://pnd.gob.mx/wp-content/uploads/2013/05/PND.pdf. Accessed 10 Apr 2017
Programa Sectorial de Medio Ambiente y Recursos Naturales 2013–2018 (PROMARNAT)
Raworth K (2012) A safe and just space for humanity: can we live within the doughnut? Oxfam international discussion paper. Oxfam International, Oxford
RSA Innovate UK (2016) Designing for a circular economy: lessons from the great recovery 2012–2016. The great recovery. Redesigning the future. ISBN:978-0-901469-85-4
Ruiz I, Barraza L, Bodenhorn B, Reyes V (2009) Evaluating the impact of an environmental education programme: an empirical study in Mexico. Environ Educ Res 15(3):371–387. https://doi.org/10.1080/13504620902906766
Sánchez M (2014) Structural challenges for the Mexican economy. The Bank of Mexico and the Hong Kong University of Science and Technology. Hong Kong, China. http://www.banxico.org.mx/dyn/publicaciones-y-discursos/discursos-y-presentaciones/discursos/%7B5E768822-DDCE-0D70-9539-ED763A2E34DC%7D.pdf. Accessed 6 Apr 2017
SEMARNAT (2012) Residuos. Informe de la Situación del Medio Ambiente en México. Compendio de Estadísticas ambientales, indicadores clave y de Desempeño Ambiental. http://apps1.semarnat.gob.mx/dgeia/informe_12/. Accessed 10 Apr 2017
Tabellini G (2010) Culture and institutions: economic development in the regions of Europe. J Eur Econ Assoc 8:677–716. https://doi.org/10.1111/j.1542-4774.2010.tb00537.x
Top Markets Report (2016) Environmental technologies a market assessment tool for U.S. exporters. U.S. Department of Commerce, International Trade Administration, Industry & Analysis (I&A). ITA 2016. http://trade.gov/topmarkets/pdf/Environmental_Technologies_Top_Markets_Report.pdf. Accessed 5 Apr 2017
Velasco MG (2011) Generation and disposition of MSW in Mexico and potential for improving waste management in Toluca Municipality. Department of Earth and Environmental Engineering Columbia University
Yuan Z, Bi J, Moriguichi Y (2008) The circular economy: a new development strategy in China. J Ind Ecol 10:4. https://doi.org/10.1162/108819806775545321

Chapter 3
Solid Waste Management for Circular Economy: Challenges and Opportunities in Romania – The Case Study of Iasi County

Cristina Ghinea and Maria Gavrilescu

Abstract In this work we proposed to answer some questions related with solid waste management and circular economy. The research questions were: What is the connection between circular economy and solid waste sector? How can the circular economy be implemented and which are the necessary policies for waste recycling? Which are the environmental issues related with the waste management in Europe? What are the main problems and opportunities in solid waste management in Romania? Can it be achieved a medium- and long-term prognosis of solid waste generation? In this paper we also suggested strategic solutions for integrated waste management and policies for its improvement. We have discussed the zero waste to landfill target together with the public perception and participation in waste management sector. We have applied the life cycle assessment for environmental evaluation of the proposed waste management scenarios and the trend analysis for waste prognosis.

Keywords Iasi · Romania · Life cycle · Public perception · Zero waste

3.1 Introduction

Waste is a resource – everyone should be aware and acknowledge this, starting from those who manufacture products and continuing with the population who consumes these products. This should be the *leitmotif* of those who are responsible to recover,

C. Ghinea (✉)
Faculty of Food Engineering, Stefan cel Mare University of Suceava, Suceava, Romania
e-mail: cristina.ghinea@fia.usv.ro

M. Gavrilescu (✉)
Faculty of Chemical Engineering and Environmental Protection, Department of Environmental Engineering and Management, "Gheorghe Asachi" Technical University of Iasi, Iasi, Romania

Academy of Romanian Scientists, Bucharest, Romania
e-mail: mgav@tuiasi.ro

M.-L. Franco-García et al. (eds.), *Towards Zero Waste*, Greening of Industry Networks Studies 6, https://doi.org/10.1007/978-3-319-92931-6_3

reuse and recycle the "resources" during the manufacturing and consumption phases. In the last years, *sustainable materials management* was promoted as an approach for the exploitation of materials in a sustainable way in order to reduce environmental impacts and preserve natural capital. It is considered that *sustainable materials management* is a precursor of circular economy, which promotes recycling, reuse and remanufacturing (ISWA 2015).

The classical model of industrial development is based on a linear model, with inputs (raw materials, energy and other resources) and outputs (wastes of products and production which are usually treated by *end-of-pipe* techniques). After ending their period of utilisation, the products are discharged on landfills or incinerated as waste. Usually this model of industrial development ("take-make-dispose") is associated to resource depletion and high environmental impacts (ISWA 2015; WEF 2014).

It was estimated that around 65 billion tons of raw materials were processed by the industrial system at the end of the first decade of the twenty-first century (in 2010), and this quantity is expected to reach about 82 billion tons in 2020 (WEF 2014). The continuous growth of the raw materials price, especially from the natural reserve, the increasing pressures of the society and regulatory constraints are connected with both resources depletion and environmental impacts generated by waste discarded in the environment.

This is why in the last two decades, *circular economy* (CE) is gaining growing global consideration as new development model able to influence the existing production and consumption model. This influence is possible based on increasing resource throughput as a condition for continuous growth (Ghisellini et al. 2015; Zils 2014).

Studies on the application of circular economy principles propose a refined hierarchy of resource use and adoption of the waste value-based recovery concept and related collection practices (Gharfalkar et al. 2015; Ghinea et al. 2012; Singh and Ordonez 2016). Therefore, new actions should be addressed in concerted ways so as to add value to the "closing the loop" concept. Intensified recycling and reuse practices along product life cycles can bring benefits for both environment and economy, with intrinsic positive impacts to the society, as at large.

3.2 Waste Management in the Context of Circular Economy

In the first part of this section, the concept of circular economy is visited in connection to the waste sector; this was with the intention to outline the importance that wastes have in the sustainable strategy like circular economy.

The materials recovery and recycling concepts were also discussed in this section in order to highlight the importance of recovery and recycling of materials and to present the recycling situation in Europe.

3.2.1 Circular Economy and the Waste Sector

There are several reasons for business to start focusing towards industrial production that tends to decouple its development from natural resources exploitation. Numerous studies and research are demonstrating that this new model of industrial development is able to foster sustainable economic growth and create new jobs. This model is based on closed loops or circular setups associated with the concept of circular economy and is materialised by new practices such as (a) reengineering/remanufacturing and (b) exploiting the whole technical and economic value of materials all along the life cycle with favourable consequences on waste minimisation by reusing, refurbishing, maintaining, recycling and recovering operations (Zils 2014).

Circular economy (CE) is a relatively simple but a sustainable strategy, which strives to reduce the inputs of virgin raw materials as well as the generation of waste by closing the economic and environmental loops of resource flows (Haas et al. 2015; MacArthur Foundation 2012). The circular economy, considered as a regenerative and restorative system, determines the shift from the *cradle-to-grave* approach to new ones such as *cradle-to-cradle* and *cradle-to-gate*, which means "closing the loop" of the life cycle of a process, product or service by recycling and reuse (Zils 2014).

According to the Circular Economy Package which includes the European legislation on waste, there are challenging targets of waste minimisation, based on long-term approaches devoted to waste management and recycling.

The key horizons of these approaches include the following targets (COM 2011):

- 2030: recycling of 65% of municipal waste
- 2030: recycling 75% of packaging waste
- 2030: reducing land filling of municipal waste by 10%
- Landfill discouraging and interdiction of landfilling of waste collected separately
- Promotion and stimulation of industrial symbiosis

Circular economy allows for improving the added value of processes and products by avoiding waste. An estimation of the European Commissions (COM 2014) reveals that the improvement of resources efficiency could reduce the entering raw materials in industrial systems by 17%–24% until 2030, which would be equivalent to €630 billion savings per year at the level of European industry, or to an increase of GDP in EU.

Maximising recycling and minimising waste, reducing natural resources consumption and reusing resources and waste are the key actions connected to "zero waste" philosophy. This is based on management principles which compete to achieve the "age of zero waste". There are members states which already apply principles of "zero waste" philosophy. The EU action plan for the circular economy stimulates the preservation of "the value of products, materials and resources in the economy for as long as possible" and waste minimisation or elimination (COM

2015). This policy can be made feasible by going through several steps, starting with tracking waste data and continuing by defining zero waste, prioritising waste-reduction tasks, strengthening supplier partnerships, resolving regulatory challenges, achieving landfill-free waste management and sharing the best practices (changing rules, creating jobs for the environment, promoting producer responsibility, recovering resources, empowering consumers, producing by cleaner technologies, designing for the environment, shifting subsidies, etc.).

The goal of recycling is not purely achieved by high recycling rates, also needs the transformation of the whole production system and adaptation of consumption via technological and social innovation (Jávor 2015).

As an example on the CEP implementation at the European glass industry, representatives of this industry proposed to the European Commission (FEVE 2015) the following actions: (a) stimulation of bottle-to-bottle multiple recycling, (b) evaluation of calculation methods and impacts of recycling targets, (c) a better definition of reuse, (d) encouraging separate collection of glass, (e) the avoidance of double targets and (f) the clarification of multi-material and municipal waste definitions. According to FEVE (2015), the EU circular economy can be made real in seven steps from the glass packaging industry view:

- *Design*: glass is designed for the environment and is 100% recyclable.
- *Production*: has increased by 39.5% in the last 25 years, and the industry maintains 125,000 direct and indirect jobs across Europe.
- *Distribution*: more than 50% of glass bottles are delivered to customers within a 300 km distance.
- *Consumption and reuse*: 87% of Europeans prefer glass.
- *Collection*: over 70% of all glass bottles are collected for recycling annually.
- *Recycling*: glass is available for multiple recycling.
- *Raw materials*: 1 tonne of recycled glass saves 1.2 tons of virgin raw materials and avoids 60% of CO_2 emissions.

Glass is a material that can be recycled indefinitely, and it is also considered that reusing glass is another way to recycle. Another sector which has increased in Europe in the last years is represented by recycling of paper and cardboard waste. The paper recycling rate in 2012 was 71.7% according to CEPI (2012).

Some necessary policies required by ERPC (2011) on paper recycling:

- *Renewable energy policy*: paper recycling must be considered a priority over energy recovery; paper can be used for incineration (energy production) at the end of life when it cannot be recycled anymore.
- *Collection of paper*: development of effective separate collection systems is necessary.
- *Trade of paper for recycling*: there are some issues related to exports of unprocessed waste paper by organisations which are not part of the recycling sector. European countries should focus on increasing the collection of paper for recycling and to maintain the collected volumes at high levels.

- *Recyclability*: policy related with the product should ensure that paper can be recycled at the end of life.

The recycled paper could replace virgin materials to produce a particular product and to reduce emissions (Ghinea et al. 2014).

An important sector of the European economy is represented by plastic industry. Plastics are valuable materials with various applications in everyday life, and they have the potential to be recycled many times (Comanita et al. 2016; PlasticsEurope 2015a). About 50% of collected plastic goes to landfilling, while only 25% is recycled (COM 2015). Plastic innovation can contribute to the circular economy by improving the plastic recyclability, by better food preserving (COM 2015). PlasticsEurope (2015a) provides some recommendations for circular economy, e.g. the recognition of resource efficiency achievements attained in the production of plastics (e.g. only 0.3% material losses in production of polypropylene), and the eco-design requirements should be based on the entire life cycle. Additionally, the legislation governing the legal status of products derived from plastic waste must be effective and consistently applied in which health and safety of workers and consumers must be taken into consideration, among other important aspects. Recycling and energy recovery are necessary to be intensified and applied in a complementary manner in order to achieve the goal of *zero plastics* to landfill by 2020 (Comanita et al. 2015). In order to obtain a recycling society, efforts are being made within and beyond Europe (Ghinea et al. 2014).

3.2.2 Materials Recovery and Recycling

Material recovery is a massive challenge with the aim to minimise the amount of waste sent to landfills by recycling and remanufacturing (Ghinea et al. 2013). The materials which can be recovered from municipal solid waste mainly include paper (basis for new paper production), glass (basis for new glass production or direct reuse of bottles), plastic (basis for new plastic production), electronic scrap and metals (recovering of gold, molybdenum, copper, etc.) and energy (from residual waste: incineration with combined heat and power – CHP) (Ghinea et al. 2013; Hall 2010).

As part of the material recovery, recycling has evident environmental benefits at every stage in the life cycle of a product from the raw material extraction to its final disposal. Recycling reduces air and water pollution associated with making new products from raw materials (such as greenhouse gases emissions, which contribute to global warming) (Ghinea and Gavrilescu 2010b; US EPA 2007). For example, according to Maguin (2015), aluminium recycling saves up to 92% of CO_2 emissions, while steel recycling saves up to 58% of CO_2 emissions.

Recycling is a complex method for protecting the environment by recovering materials or components of used products, resulting in new products (Ghinea 2012). The purpose of recycling is to limit the use of new materials and decrease the amount of waste. Some benefits of recycling are resource conservation (reducing demand for new resources), reducing transport costs and energy production (energy saving by avoiding exploitation of raw materials) and saving resources that otherwise would be lost in storage sites (Ghinea 2012).

Recycling is preferred to incineration and disposal of non-renewable materials such as glass, metals and plastics because the total energy and global warming potential is generally low. For renewable materials (paper and cardboard) in most cases, the global warming potential is lower for recycling than for incineration. Recycling can be sustainable if it is efficient in terms of costs (Ghinea 2012).

European countries have made considerable efforts to convert waste into resources and to promote recycling as a sustainable waste management stage. In 2014 at European level, 7.7 million tonnes of plastic waste were recycled. Plastic demand in Europe in 2014 was 39.5% packaging, 20.1% building and construction, 8.6% automotive, 5.7% electrical and electronic, 3.4% agriculture and 22.7% others (PlasticsEurope 2015b). In 2014, 25.8 million tonnes of post-consumer plastics waste were generated from which 29.7% was recovered through recycling and 39.5% by energy recovery while 30.8% was sent to landfill. According to PlaticsEurope (2015b), the recycling percentage has increased from 2006 to 2014 with +64%. The landfill of plastic is still the first option in many European countries. For plastic waste the recycling option is preferred, and energy recovery is the alternative for plastics which cannot be more environmentally friendly recycled (PlasticsEurope 2015b).

About 58 million tons of paper waste were recycled in Europe in 2014, which is almost 71.7% of all paper consumed (ERPC 2015). According to ERPC (2015), Europe continues to be the world leader in paper recycling, followed by North America.

In Europe according to FERVER (2015), the glass production exceeds 30 million tons per year, and 70% of the material is recycled. The glass can be recycled without loss of material; sorting of waste glass plays an important role in recycling. Glass recycling reduces energy and CO_2 emissions in glass production and creates business and jobs. Two key factors in building a sustainable circular economy are efficient use of raw materials and recycling of packaging in a closed loop, the glass bottle to bottle recycling initiative (FEVE 2015).

However, Eurostat statistics on waste reveals that each European citizen generated 475 kg of waste in 2014. From this amount, only 44% is being recycled or composted, while the remaining 56% ended up landfilled (28%) or incinerated (27%). Zero Waste Europe's Executive Director, Joan Marc Simon said "A residual waste target of 100 kg per capita for 2030 is a good indicator of resource efficiency and resource use, as it works on the top levels of the waste hierarchy, effectively combining prevention, reuse and recycling policies" (ZWE 2016). In this context, the statistics show that there is a large variability across Europe regarding waste generation and waste treatment (ZWE 2016).

Moreover, waste generation in some member states like Romania, Poland or Latvia is under the EU average of 259 kg per inhabitant with less than 300 kg per inhabitant, while some others generate considerably more than the EU average, being over 600 kg per inhabitant (Cyprus and Germany, among others) and even over 750 kg per inhabitant (Denmark). Even further, Slovenia, Romania and Poland are among the recently joined members with the lowest rate of waste generation (under 200 kg per inhabitants). Although with the highest rate of waste generation in Europe, Denmark is the member state with the lowest rate of waste landfilling, together with Germany, Belgium, the Netherlands and Sweden. From Eurostat statistics, Slovenia is the best performing EU country having the lowest amount of waste landfilled (102 kg per inhabitant in 2014) (ZWE 2016).

Therefore, zero waste in Europe in the context of circular economy needs more concerns in terms of waste generation improvements as well as regarding increasing the extent of waste management by recycling.

3.3 Environmental Issues, Resource Conservation and Drivers in Waste Management in Europe

Nowadays, the human ecological footprint overshoots Earth's carrying capacity (Best et al. 2008; Gavrilescu 2011; Holmberg et al. 1999). Resource consumption and production of higher levels of greenhouse gases, toxics and nondegradable waste are higher than this planet has the capacity and the time to regenerate or assimilate. These are the most important issues that must be considered by the entire humanity in relation with sustainability. The resource productivity, dematerialisation and a change from efficiency thinking to effectiveness and sufficiency are required. This latter is in relationship to the fact that materials demand is expected to grow in the following decades (Rossy et al. 2010). In this context the *sustainable materials management* (SMM) concept represents one of the frameworks to sustain strategies for the future and sustainable development visions in the context of circular economy. Hence, the SMM definition as published by the OECD is as follows: *sustainable materials management* is an approach "to promote sustainable materials use, integrating actions targeted at reducing negative environmental impacts and preserving natural capital throughout the life-cycle of materials, taking into account economic efficiency and social equity" (OECD 2007). SMM supports a life cycle view of impacts associated with material use and strongly encourages the creation and execution of policies and practices relating to materials. Future policy directions for SMM are transition from eco-efficiency to eco-effectiveness and further to conservation, to eco-sufficiency (OECD 2011a; Rossy et al. 2010). It can be considered that SMM is a precursor of circular economy.

SMM acts to support sustainable development by a shift from waste management to materials management (OECD 2011a). Moreover, the OECD issued the policy principles of the SMM (OECD 2011a; Rossy et al. 2010; Wante 2010) at different levels, i.e. (i) to preserve natural capital; (ii) to design and manage materials, products and processes for safety and sustainability from a life cycle perspective – maximising positive and minimising negative impacts to the environment and human health/well-being through eco-design; (iii) to use the full diversity of policy instruments to stimulate and reinforce sustainable economic, environmental and social outcomes; and (iv) to engage all parts of society to take active, ethically based responsibility for achieving sustainable outcomes.

The above-mentioned SMM policy principles (SMMPP) have been applied in different countries (AG 2009; OECD 2011a; SEPA 2008; SME 2003) such as Sweden, Finland, Belgium, Czech Republic and the Netherlands. A brief description of those policies is described in the following paragraphs:

- Sweden and Finland, as representatives of some Scandinavian countries, have the *natural capital preservation* in the core of their SMMPP. Sweden established 16 environmental quality objectives so as "to promote human health, safeguard biodiversity and the natural environment, preserve the cultural environment and cultural heritage, maintain long-term ecosystem productivity and ensure wise management of natural resources", while Finland developed long-term goals on how to use natural resources sustainably and to improve eco-efficiency.
- In the case of Belgium, it has developed various strategies such as the transition network established by the Belgian Public Waste Authority in order to promote SMM. Also, Belgium is focused on identifying material cycles with high potential for reduced environmental impact and on collaboration of different actors in order to set up policy measures that work in different phases of the materials life cycle.
- In Czech Republic aside of a large variety of policies and policy instruments including regulations, economic incentives and disincentives, trade and innovation policies, information sharing and partnerships were developed to promote sustainable development. About information sharing the Czech Environmental Information Agency allows public access to different sources of information and statistics on many aspects of sustainability and the environment.
- The Netherlands has one of the best practices in Europe regarding recycling of paper/cardboard among other materials and has a leading position promoting circular economy, globally. In order to achieve SMM, as part of the national Future Waste Policy, the Netherlands has adopted a "Material Chain Approach" (Dutch Chain Approach) which represents a general movement towards SMM. During the implementation of the Dutch SMM approach, the government's role is rather as convener and facilitator (LAP 2 2008; OECD 2011a). The Dutch government ambitions are to minimise environmental pressure over the whole supply chain and to harmonise policy in different areas by means of a chain-oriented waste policy (LAP 2 2008; OECD 2011b; Veeken et al. 2011). The overall goal is to reduce the environmental impact of material chains throughout the life cycle in the most cost-effective manner (LAP 2 2008; OECD 2011b).

Even beyond the European borders, there are other countries framing their waste management under similar policies as the SMMPP: one of them is Australia which has drafted a National Action Plan for Education for Sustainability. The actual name of such plan is "Living Sustainably" that stimulates collaboration between representatives from academia, non-government organisations, youth and local government.

Improving waste management is known as an environmental challenge at international level. The implementation plan agreed at the World Summit on Sustainable Development (WSSD) (Johannesburg, September 2002) was built on the Agenda 21 and calls for further action to "prevent and minimise waste and maximise reuse, recycling and use of environmentally friendly alternative materials, with the participation of government authorities and all stakeholders, in order to minimise adverse effects on the environment and improve resource efficiency" (EC 2005).

In 2005, the European Commission adopted the *Thematic Strategy on the prevention and recycling of waste* which addresses waste prevention as one of the priority issues (EC 2005). This Strategy is one of the seven thematic strategies contained by the 6th Environment Action Plan, which "aims to help Europe become a recycling society that seeks to avoid waste and uses waste as resource" (EC 2016). It entails the use of economic instruments to implement the waste hierarchy so that key actions have to be set out to modernise the existing legal framework and to encourage waste prevention, reuse and recycling, with waste disposal only as a last option (EC 2011a; JRC 2011).

In the context of a continued growth of the world population, combined with stronger emerging economies, the *Thematic Strategy on the prevention and recycling of waste* has "to adapt the EU waste policy approach to this new reality" taking into account the significant increase in the total consumption (DEFRA 2011; EC 2011a; OECD 2010; von Braun 2008). Legislative improvements are the first step towards adapting the regulatory framework, in order to provide a legal structure that is flexible and promotes a recycling society, which avoids losses and uses resources that are found in waste (EC 2005).

Further developing approaches are necessary for determination of the suitable options from technical, economical, social and environmental points of view. Also, setting targets for waste recycling and recovery must be performed taking into account the differences between products and materials and possible alternatives.

The development of innovative policies for sustainability in the medium and long term needs the constant tuning between actions of private operators and the regulatory frame. As a consequence, the spectrum of existing and new waste treatment technologies and managerial strategies has also shifted from maintaining environmental quality at present to meet sustainability goals in the future.

The Waste Directive 2008/98/EC (EC Council Directive 2008) reflects engagement of the EU towards sustainable development, in particular to the SWM systems. Also, it brings new challenges, such as new definitions for waste, by-products and end-of-waste criteria, resulting in the need to (1) choose appropriate technologies that aim at improving the protection of human health and environment, (2) promote reuse and recycling, (3) enhance waste prevention programmes via biowaste separate collection and (4) implement extended producer responsibility (EPR) collectively (Chang et al. 2011, Pires et al. 2011). In addition, key challenges related to long-term waste management are climate change and energy use link SWM systems to the reduction of greenhouse gas (GHG) emissions and the enhancement of energy recovery.

As shown previously, one of the most relevant objectives is to reduce the amount of waste generated. However, these efforts are still very limited, especially in some south-eastern European countries, like Romania, with mixed results and relatively few efforts that have been made to regulate the management of various categories of waste. One of them are the organic materials that usually comprise over 50% of the total waste generation in the cities, construction and demolition waste (Bayer and Méry 2009; Björklund and Finnveden 2005; Papachristou et al. 2009).

3.4 Solid Waste Management in Romania and Iasi County Case Study

In the following sections, we present and discuss some challenging issues and opportunities concerning solid waste management in Romania, taking Iasi County as case study. Separate collection of waste fractions in solid waste is treated as the key factor which we consider as a condition to promote and implement management practices as reuse, recycling of waste in Romania. This approach is adopted since landfilling continues to be the main treatment/elimination method for solid waste – known as the most unfavourable alternative according to the waste hierarchy and totally opposite to what the circular economy promotes. Some policies on waste management in Romania are also presented, along with information on quantities of waste generated in Iasi city and waste amounts forecasting on short, medium and long terms. A trend analysis of solid waste prognosis is also illustrated together with data on mixed collection and waste landfilling. In Sect. 3.5 of the following chapter, we propose some waste management scenarios as sustainable integrated alternatives for solid waste management in Iasi. These were evaluated from an environmental impacts perspective, by applying life cycle assessment. After this evaluation, we established the most suitable scenario from an environmental point of view. This scenario includes unit operation that includes the following steps: *separate collection by waste fraction*, *recycling of materials*, *composting organic waste* and *incineration of the residual waste*. We also make a comparison with the Dutch waste management approach. We found that the Dutch waste management system can be used as a model for waste management in Romania at this moment, even if the Dutch waste management system needs to improve certain aspects. In Sect. 3.6 we present some information about circular economy, policy, proposals for improvement and some aspects regarding material recycling in Romania. We also discuss the possibility to implement of zero waste to landfill in Romania and the public perception and participation in separate collection of waste.

3.4.1 *Romanian National Strategy and Policy on Solid Waste Management: Critical Issues and Targets*

Romania is a European country situated in the south-eastern side of Central Europe, north of the Balkan Peninsula. The whole inferior course of the River Danube is located on its territory. Romania is a Carpathian country, bordering the Black Sea.

In terms of territorial administration and division, Romania includes 41 counties grouped in 7 regions, while the eighth region is represented by Bucharest and the county of Ilfov (Davidescu and Strat 2014; Ghinea 2012). In 2008, Romania had a population of about 21.5 million inhabitants, while just 20 million of inhabitants were reported according to the INS census in 2011, and various demographical analyses show that a further decrease in population number is expected (Eurostat 2015). Iasi County belongs to the Region 1 North East which also includes the following counties: Botosani, Suceava, Neamt, Bacau and Vaslui.

Before 1990 the waste management in Romania was hardly considered, and the first statistics were only introduced in 1993 (Schiopu et al. 2007). In 1998 the most predominant method of waste disposal was landfilling (99%), and around 250 non-compliant landfills were operational at that time in the country (Oroian et al. 2009; Schiopu et al. 2007). The first National Waste Management Strategy (NWMS) has been elaborated in 2004 for the period 2003–2013, 3 years before Romania's accession to the European Union. NWMS was established for setting waste management objectives aiming at creating the framework for developing and implementing an integrated waste management system (NWMS 2004). The National Waste Management Plan (NWMP) was developed for the implementation of NWMS, while the European acquis on waste has been transposed into Romanian legislation. Also, regional plans for waste management in 2005–2006, county plans waste management during 2007–2009, "master plans" and feasibility studies were developed for assisting and ensuring the implementation of integrated waste management systems (IWMS). This latter was based on the *Sectoral Operational Programme Environment* (2007–2013) provisions and funds. NWMP priorities were established to increase the competitiveness of the productive sector and its attractiveness to foreign investors; improve and develop the transport infrastructure, energy and environmental protection; develop human resources, increase the employment rate and combat any social exclusion; diversify the rural economy and expand agricultural productivity; and to promote balanced participation of all regions in the economic development of the country. National policy on waste management must guarantee European policy objectives in terms of waste prevention and aim to reduce resource consumption and practical application of the new approaches on waste hierarchy. The principle of preventive action is the essential one in Emergency Ordinance 195/2005 (GEO 2005) on environmental protection further amended and supplemented, while the provisions of Directive 2008/98/EC are transposed into the national legislation by Law 211/2011 regarding waste regime.

Romania implemented the *acquis communautaire* at the moment of its joining the European Union (EU) on 1 January 2007, with a few exceptions, among which we mention here those related to waste management (EC 1994, MECC 2016):

- Directive 94/62/EC on packaging and packaging waste, for which was requested and obtained a transitional period until 2013
- Directive 99/31/EC on the landfill of waste, for which was requested and obtained a transitional period until 2017
- Council Directive 2000/76/EC on waste incineration, for which was requested and obtained a transitional period until 2008
- Council Directive 2002/96/EC on waste electrical and electronic equipment (WEEE), for which was requested and obtained a transitional period until 2008
- Regulation no. 259/93 (EEC 1993) on the import, export and transit of waste by the end of 2015

The second National Waste Management Strategy for 2014–2020 was approved in 2013 by Government Decision no. 870 of 06.11/2013 (GD 2013). The need to review the first National Waste Management Strategy derived mainly from the following reasons (NWMS 2013):

- New concepts were established at the level of the European waste management approach (mainly according to the need to address waste as a resource and the extended producer responsibility principle).
- Adoption of Directive 2008/98/EC of the European Parliament and of the Council from the 19 November 2008 on waste and repealing certain directives (new precise drawings Waste Framework), its transposition into national law and the need to integrate the principles and provisions of these documents in national programming. The implementation of Directive 2008/98/EC requires our country to adopt ambitious goals in waste management such as recycling 50% of household waste and 70% of construction and demolition waste by 2020. Also, EU legislation requires the fulfilment of the objectives addressing recycling/packaging materials (glass, metals, plastics), as well as valorisation of biodegradable waste instead of its landfilling.
- Incorporating provisions and legislative requirements arisen during 2004–2012.
- Implementation of integrated waste management development projects in various stages of execution, implementing the proposed new technologies and new waste treatment in Romania.
- Institutional and organisational changes during 2004–2012.

The new NWMS aims also to create a national planning framework needed to develop and implement integrated sustainable waste management. Eight strategic objectives have been formulated to achieve the goals of NWMS (2013), namely, improvement of the environment and protecting human health, supporting research and development in waste management, encouraging green investments, increasing resource efficiency, sustainable waste management, correlation of waste management policies with climate change, development of responsible behaviour on the prevention of waste generation and management and the strengthening of institutional capacity.

In the National Strategy for Sustainable Development of Romania (2013–2020–2030), the development of integrated waste management systems is included as objective (NSDSR 2008) with the following targets:

- The annual quantity of biodegradable waste landfilled should have been decreased to 2.4 million tonnes by 2013, representing 50% of the total amount generated in 1995.
- Practices for materials recovery from packaging waste for recycling or incineration with energy recovery of 60% for paper/cardboard, 22.5% for plastics, 60% for glass, 50% for metals and 15% for wood before 2013.
- Reducing the number of historically contaminated sites in minimum 30 counties before 2015.
- Establishing 30 integrated waste management systems at regional/county level and closure of 1500 small landfills located in rural areas and the 150 oldest waste dumps in urban areas, before 2015.

In 2010, 38% of municipal waste were landfilled, 22% incinerated, 25% recycled and 15% composted in the EU (Eurostat 2010). In Romania, although significant efforts and investments have been made to align the national regulations to the EU waste *acquis*, some old waste management practices remain prevalent: waste disposing and landfilling (NWMS 2013). Although Romania is an EU member state, and the European policy in the area of solid waste management should be implemented as soon as possible, these changes are still difficult to be made.

3.4.2 Strengths and Weaknesses of Solid Waste Management in Romania

Today, in Romania the MSW management is carried out under the European regulation in the field, 100% legally implemented, and based on some documents derived from this (Petrescu et al. 2010). Waste management in Romania is characterised by the continued growth of waste quantities; inadequate waste collection (mixed collection and not selective) and economic benefits of waste are insufficiently developed (Petrescu et al. 2010).

According to NWMS (2013), the strengths of solid waste management are:

- *Legislative*: the existence of a frame harmonised with the European one
- *Planning*: existence of planning documents on all three levels (national, regional and county)
- *Human resources*: experience in development and implementation projects and in running awareness programmes
- *Financial*: 1.167 million euro for development of integrated waste management systems and rehabilitation of historically contaminated sites

Some additional strengths can be highlighted (Luca and Ioan 2014): existence of programmes to collect plastic waste from nature, existence of collection points by purchasing from inhabitants the beverage bottles in the major cities, waste paper is an easily recycled waste, possibility of involvement of local authorities in public waste management, etc.

Some weaknesses in waste management were presented in NWMS (2013) and grouped as:

- *Legislative*: frequent amending and updating legislation.
- *Planning*: insufficient integration with other plans and programmes including other institutions which develop strategies.
- *Data management*: inadequate practice in collecting, integrating and evaluating available data.
- *Implementation*: (a) failure of parties involved in taking ownership of responsibilities, (b) service coverage and low degree of expansion of separate waste collection and (c) *infrastructure* – not enough developed, including infrastructure support (road, water supply, sewerage, gas, etc.).
- *Economic*: stream recycling is still not stabilised due to insufficient implementation of separate collection, low expansion of separate waste collection systems.

Some weaknesses of waste management can be mentioned (Luca and Ioan 2014): paper is recycled to a very small extent, lack of incentive policies for selective collection of plastic waste, metal wastes which are not collected separately are difficult to recycle, selective collection of waste is not done sufficiently, etc.

3.4.3 Data About Iasi, Romania

Iasi County, situated in north-eastern Romania, had 826.552 inhabitants in 2008, with a density of 151 inhabitants per km^2. In the same year, the urban population in Iasi City was 391.654 inhabitants according to INS (2011). In the urban area, the number of inhabitants decreased during 2010–2011 (382,000 inhabitants in 2011) and is increasing from 2012 to 398.000 inhabitants (INS 2015). The relief is characterised by a high unit (with an average altitude of 300–350 metres) and a low plain (100–150 m) (Doba et al. 2008). In the west and south, the average annual temperature of air is between 8 °C and 9 °C, while in the north and north-east, it is between 9 °C and 10 °C.

In 2014 the population from Iasi County was involved in the following main activities: agriculture, forestry and fishing, industry (extractive, manufacturing, production and supply of electricity, gas, steam and air conditioning, water supply, sewerage, waste management and remediation activities, etc.), wholesale and retail trade, repair of motor vehicles and motorcycles (INS 2016). In this county there are concerns about the wrong development of organic farming by using bio-fertiliser and organic fertilisers. Animal husbandry is also an important sector with a rich tradition in this area (Doba et al. 2008; Iasi County Council 2009; INS 2011; RWMP-NE 2006).

3.4.4 *Quantities of Municipal Waste Generated in Iasi and Prognosis on Medium and Long Terms*

Municipal solid waste includes the waste generated and collected (mixed or selective), as well as uncollected waste. The amounts of waste generated in Iasi during 2009–2013 are illustrated in Fig. 3.1. These amounts decreased in 2011, increasing again after this year and reaching approximately 250,000 t/year of waste generated in 2013 (EPAIS 2014a, b). This growth in the volume of generated municipal waste is the consequence of an increased consumption of the population and a higher segment of population assisted by public health services in a centralised system (Doba et al. 2008). The household wastes represent an important component of municipal solid waste (MSW). For example, 67.73% from the total amount of MSW generated in 2012 (almost 233,000 tons) is represented by the household and assimilable waste, followed by demolition and construction waste (30.44%) and waste from municipal services (1.83%) (EPAIS 2014a). The household waste composition illustrated in Fig. 3.2 was estimated using data from the annual statistical survey questionnaires completed by sanitation operators and recyclable waste collectors (EPAIS 2014a, b) and not determined by direct measurements.

Similar to the situation at national level, the biodegradable waste represents an important component of municipal waste in Iasi County (EPAIS 2014b). From Fig. 3.2 it can be observed that biodegradable waste has the highest percentage in the total waste amount. This waste category includes biodegradable gardens and parks waste; food and kitchen waste from households, restaurants, caterers or retail stores; and comparable waste from food processing plants. According to EPAIS (2014a), the biodegradable municipal solid waste percentage dropped from 64% in 1998 to approximately 48% in 2012. Analysing the amount of waste generated in the Iasi County during 2009–2013, it can be perceived that the quantities of waste generated had a slight fluctuation and overall showed a decline. We have considered

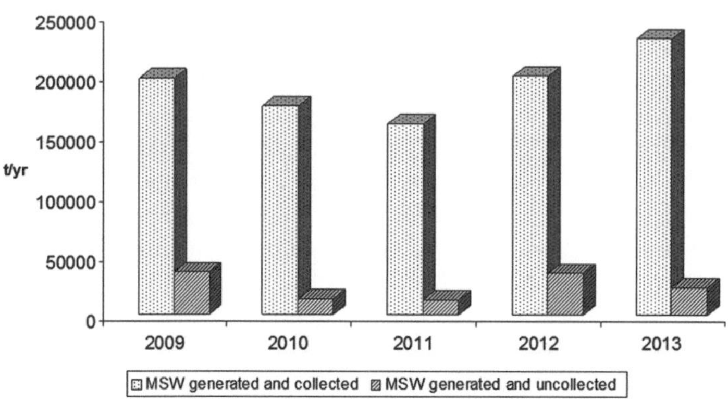

Fig. 3.1 Municipal solid waste (MSW) quantities generated in Iasi (2009–2013). (Data from Iasi County Council 2009, processed by authors)

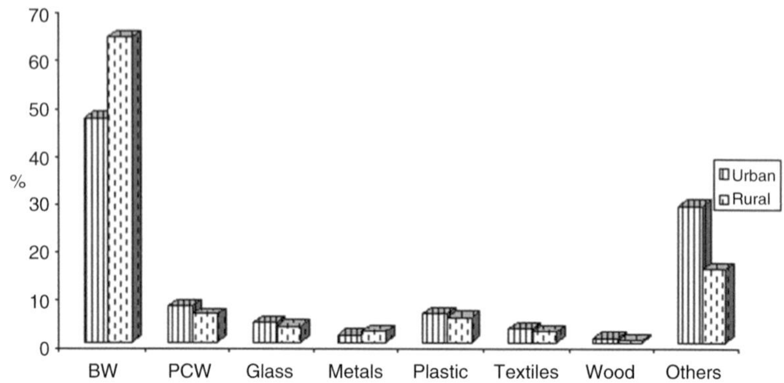

Fig. 3.2 Waste composition in urban and rural areas of Iasi county (*BW* biodegradable waste, *PCW* paper and cardboard waste). (Data from Iasi County Council 2009, processed by authors)

that this trend is mainly due to the economic crisis and decline and less due to preventive measures.

Prognosis of municipal solid waste generation has an important role in planning and implementing of waste management systems (Ghinea and Gavrilescu 2010a). Ghinea (2012) used Waste Prognostic Tool to forecast the amount of waste that will be generated in Iasi city in 2018 based on the quantities of waste generated in 2008 and on socio-economic conditions. Ghinea et al. (2016a) applied regression analysis and time series analysis included in Minitab software to forecast municipal solid waste generation and composition for the same city from Romania. According to EPAIS (2014a), the main factors that influence waste prognosis are changes in population in the county, changes in the county's economy, changes in the demand and nature of consumer goods and changes in production technologies.

The trend analysis can be also used for solid waste prognosis. In order to achieve waste prediction, we have used Minitab 17 software which includes this analysis. In this paper we will present only some of the obtained results. We have chosen to illustrate the results obtained for biodegradable and paper waste since the fraction with the highest percentage of the total quantity of waste generated is represented by biodegradable waste, while paper wastes are the most separately collected fraction. Figure 3.3a, b illustrates the S-curve trend model applied for biodegradable waste, and paper waste prognosis for Iasi, which included three variables: actual (•), fits (■) and forecast (♦) versus time. The forecasting values were sets until 2030.

The graphs provide the values for mean absolute percentage error (MAPE), mean absolute deviation (MAD) and mean squared deviation (MSD). It is considered that lower values for MAPE, MAD and MSD indicate a better fitting model. In our case we have lower values for MAPE for both waste fractions. The MAD value is lower for paper waste than those for biodegradable waste, while for MSD are registered higher values. It can be concluded that for the paper waste fraction was obtained a better fitting model. The form of graphs is similar for both fractions. The prediction equations are also presented in the graphs. We will use the data obtained

Fig. 3.3 S-curve trend model for: (a) biodegradable waste, (b) paper waste (authors' Figures)

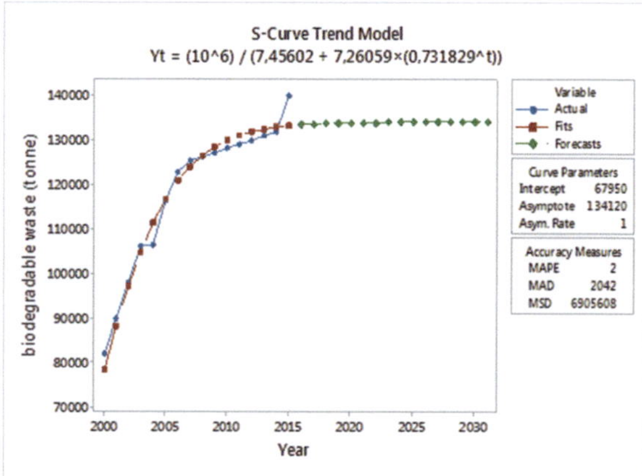

a)

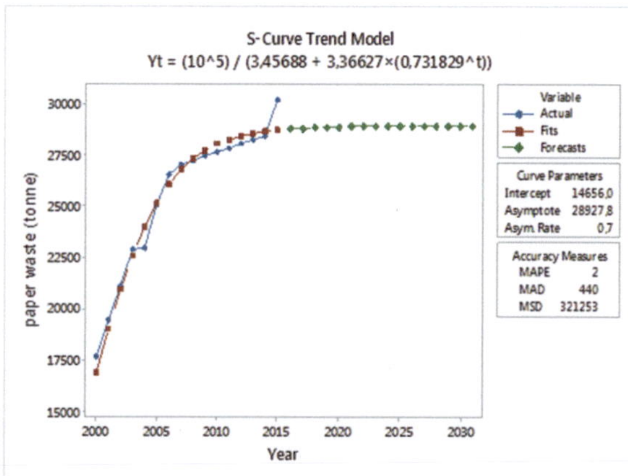

b)

from the prognosis analysis to propose waste treatment methods which can be included then in different waste management systems and evaluated with various methodologies in order to establish the most suitable MSW system for implementation.

3.4.5 Waste Collection, Transportation and Separation at Source

The municipal solid waste is collected by a public company of local interest, SC Salubris S.A. Iasi. The separate collection of waste is performed with small steps, in pilot projects, usually for materials with high market values. In general, the inhabitants use to bring the solid waste to the collection points distributed over the town at

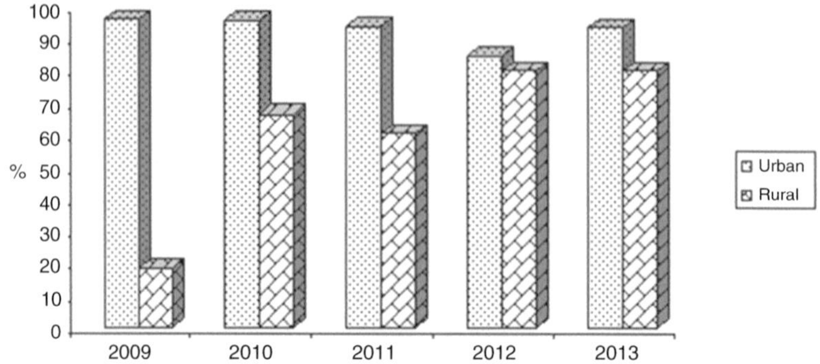

Fig. 3.4 Evolution of coverage degree with sanitation services. (Adapted from EPAIS 2014a)

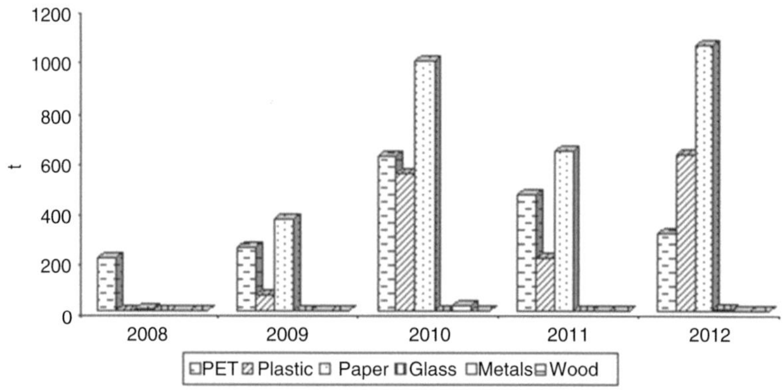

Fig. 3.5 Quantities of waste from households collected selectively in 2008–2012. (Adapted from EPAIS 2014b)

special locations, close to their housing, and place the waste in containers provided by the waste collection company. For separate collection, there are special containers for waste fractions in some collection points located in various parts of the town (Doba et al. 2008; Ghinea 2012). The coverage degree with sanitation services for both urban and rural areas is presented in Fig. 3.4, which shows that rural areas are also covered. The amounts of households waste collected selectively in 2008–2012 are presented in Fig. 3.5 (EPAIS 2014a).

Unfortunately, this practice to collect the mixture of waste without separation at the source is common at national level. Three stages for implementation of selective collection were proposed (BALKWASTE 2010): pilot projects and public awareness (2004–2006), selective expansion of collection at national level (2007–2017) and implementation of selective collection in more difficult areas (2017–2022). Selective waste collection is still in its infancy, while more informative campaigns for citizens to separately collect a larger amount of waste are necessary. The informative campaigns are taken from the example of other EU countries, which purpose is to help to protect the environment and human health.

If until 2009 the collection of waste was performed mostly in urban areas, after 2009, waste collection was expanded also in rural area (Fig. 3.4). It can be observed that collection of waste generated was significantly improved, and this aspect is important because efforts were and are made to collect all amounts of waste generated, so no longer remain wastes dumped in different areas. From Fig. 3.5 it can be observed that was recorded a small increase in the amounts of different waste fraction collected selectively. Paper and plastics are the most selectively collected waste fractions. Considering the data from these figures, we can say that improvements of waste management system should be proposed, focusing mainly on selective collection, recovery and recycling.

3.4.6 Waste Treatment and Landfilling

In 2008 the municipal solid waste management (MSWM) system in Iasi included only waste landfilling at the Tomesti landfill. This was a significant source of soil and groundwater pollution because the leachate collection network was totally improper. The old Tomesti landfill was closed in 2009, when a new compliant landfill was built and put into operation in accordance with the legislation, situated at almost 8 km from the city of Iasi, in Tutora centre for integrated waste management. Tutora landfill surface area is 50 ha, with a storage capacity of 8,613,000 m^3 and with four cells that should serve the entire Iasi County. It includes a leachate collection system, a landfill gas collection system and a rainwater collection system. The leachate treatment plant uses the reverse osmosis (RO) technology with a processing capacity of 84 m^3 leachate/day. The RO plant is complying with the requirements of EU directives (EC Council Directive 1999; EC Council Directive 2008). The centre for integrated waste management Tutora includes, in addition to the landfill, a sorting station with a capacity of 29,000 tons/year and a composting station. This latter was put under construction in 2009. In March 2012 the composting process was supposed to start, based on a turned windrows method, with a pyramidal shape. Since October 2012 windrows with green waste and household waste (25–30%) were supposed to be carried out, but because of various reasons, the composting station did not work in 2012 and 2013 (EPAIS 2014a, b).

3.5 Waste Life Cycle and Strategic Solutions for Integrated Waste Management Systems

In this section we presented eight municipal solid waste management scenarios developed by us, which were evaluated from environmental point of view in order to establish the most suitable scenario for implementation. The evaluation was performed using life cycle assessment (LCA) method. A hierarchy of environmental impacts resulted considering different LCA methodologies are also presented. In this section, a comparison with waste management system from Twente region of the Netherlands is performed to emphasise that systems like the one in the Netherlands could serve as models.

3.5.1 Development of Municipal Solid Waste Management Scenarios for Evaluation and Implementation in Iasi County

Considering specific and relevant methods for treatment of municipal solid waste, we planned and proposed different alternatives for MSW management system for Iasi, Romania (Ghinea 2012; Ghinea et al. 2012). Eight scenarios were developed in order to evaluate them from an environmental point of view (Ghinea 2012). The scenarios labelled from **2 to 8** represent integrated alternatives of waste management unit operations including various methods for treatment of waste. Nevertheless, all scenarios included temporary storage of waste in containers, their collection and transport. The differences between these scenarios consist in the treatment or elimination method chosen (Fig. 3.6) (Ghinea 2012):

- **The first scenario** (**S1**) represents the *system that was in operation until 2009*. Municipal solid waste was landfilled in Tomesti, a non-compliant landfill without systems for collection and treatment of landfill gas and leachate (Doba et al. 2008).
- **The second scenario** (**S2**) *includes mixed waste collection, landfilling with collection of biogas and treatment of leachate* on the new landfill that was established in 2009. Facilities compiled with the landfill directive provisions (EC Council Directive 1999).
- **The third scenario 3** (**S3**) includes composting as a method of treating biodegradable waste. The resulted compost can be used in agriculture to replace the synthetic fertilisers. Also, Scenario 3 includes landfilling of municipal solid waste with leachate treatment and biogas collection and valorisation.
- **The fourth scenario** (**S4**) includes both composting and landfilling as solid waste treatment methods but also sorting of recyclable materials.
- **The fifth scenario** (**S5**) includes sorting, composting, landfilling and incineration as a method of treating residual waste in order to generate electricity, with the metal recovery from the treatment of slag and landfilling of ash obtained.
- In the **sixth scenario** (**S6**), there are two ways of dealing with biodegradable waste such as composting (aerobic process) and anaerobic digestion from which compost and biogas result. Biogas can be used as a source of energy.
- **The seventh** (**S7**) and **eighth** (**S8**) **scenarios** include composting of biodegradable waste and waste incineration. The main difference between them is that S7 includes sorting of recyclable materials.

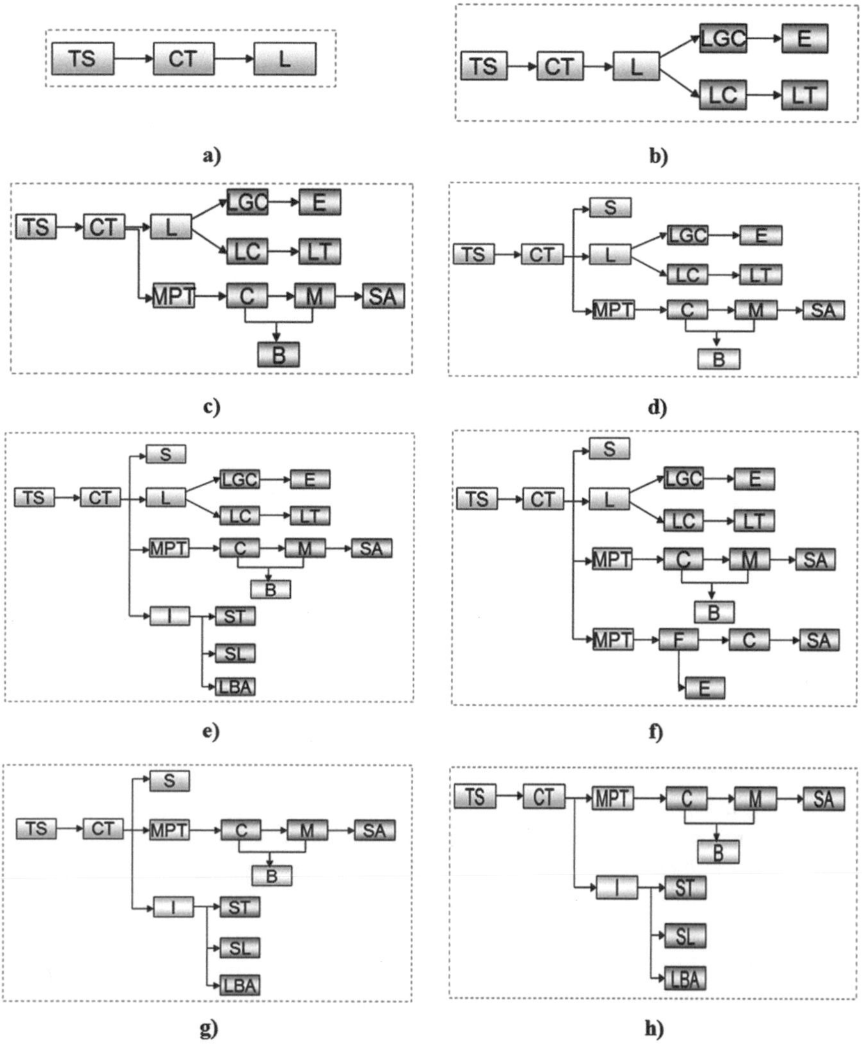

Fig. 3.6 Scenarios proposed for the analysis of MSW management in Iasi, Romania: (**a**) S1, (**b**) S2, (**c**) S3, (**d**) S4, (**e**) S5, (**f**) S6, (**g**) S7, (**h**) S8; *TS* temporary storage, *CT* collection and transport, *L* landfilling, *LGC* landfill gas collection, *LC* leachate collection, *E* engine, *LT* leachate treatment, *MT* mechanical treatment, *C* composting, *M* maturing, *SA* soil application, *B* bio-filter, *S* sorting, *I* incineration, *ST* slag treatment, *SL* slag landfilling, *LBA* landfilling bottom ash, *F* fermentation

The collection of waste is separate for biodegradable waste and mixed for the other waste fractions (in the case of scenarios 3 and 8). For scenarios 4–7 it was assumed that the waste fractions are collected separately.

All scenarios were evaluated from an environmental point of view by applying life cycle assessment (LCA) methodology, according to its four steps: goal and scope analysis, inventory analysis, impact assessment and interpretation. The scope

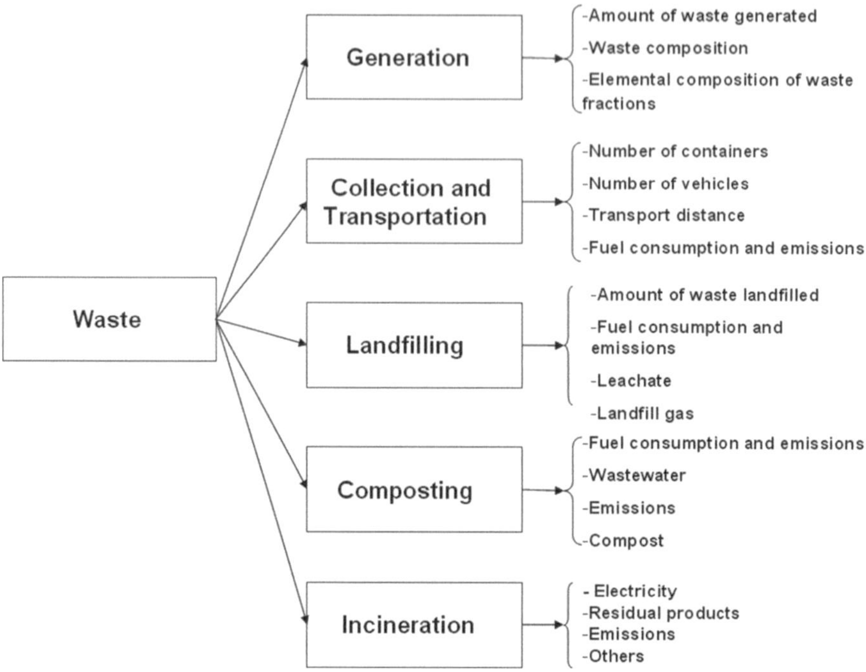

Fig. 3.7 Waste characteristics considered for evaluation

of the study was to determine the environmental impacts of all scenarios proposed and to establish the most suitable scenario for implementation considering the environmental aspects. In the inventory analysis phase, all the inputs and outputs were established for each step included in the scenarios. The characteristics considered for the evaluation are illustrated in Fig. 3.7.

The impact assessment stage was performed using GaBi software, one of the most popular software tools for LCA. After entering all inputs and outputs for each process and connecting the processes, GaBi software allowed us the calculation of the material and energy balances in order to assess the environmental impacts. The evaluation was performed using some of the LCA methodologies included in GaBi software: CML 2001, CML 1996, EDIP1997, EDIP2003, EI95, EI99, etc. CML 2001 is an environmental themes method developed by the Centre of Environmental Science of Leiden University and succeeds the CML 1996 methodology. CML 2001 methodology includes a set of impact categories defined for the problem-oriented approach (Frischknecht et al. 2007; JRC European Commission 2010). Environmental design of industrial products (EDIP 1997) is a Danish LCA methodology, a documented midpoint approach covering most of the emission-related impacts, resource use and working environmental impacts (Wenzel et al. 1997). EDIP 2003 is an update of the EDIP 1997 methodology and a theme-oriented method. The main difference between EDIP 1997 and EDIP2003 lies in the selection of the category indicator (Hauschild and Potting 2005). In the Eco-Indicator 95 (EI 95) methodol-

Table 3.1 The hierarchy of environmental impacts resulting from application of the methodology CML 2001

Impact category	Environmental impacts of scenarios negative impacts → positive impacts
Global warming potential (GWP)	S1 > S2 > S3 > S8 > S5 > S6 > S4 > **S7**
Eutrophication potential (EP)	S1 > S5 > S6 > S7 > S4 > S3 > S2 > S8
Photochemical ozone creation potential (POCP)	S1 > S3 > S2 > S5 > S6 > S4 > S8 > **S7**
Acidification potential (AP)	S1 > S5 > S6 > S3 > S4 > S2 > S8 > **S7**
Human toxicity potential (HTP)	S2 > S1 > S3 > S6 > S4 > S5 > S8 > **S7**

Table 3.2 The hierarchy of environmental impacts resulting from application of CML96, EDIP1997, EDIP2003, EI95, EI99 and HA methodologies

LCA methodology	Environmental impacts of scenarios positive impacts → negative impacts
CML96	**S7** > S6 > S8 > S5 > S4 > S3 > S2/> S1
EDIP1997	S8 > **S7** > S2 > S4 > S6 > S5 > S3/> S1
EDIP2003	S8 > **S7** > S4 > S6 > S5 > S2 > S3/> S1
EI95	S6 > S4 > S3>/ S1 > S5 > S7 > S8 > S2
EI99, HA	**S7** > S4 > S6 > S3 > S5 > S8 > S2/> S1

ogy, "the environmental effects (greenhouse effect, ozone layer depletion, acidification, eutrophication, smog and toxic substances) that damage ecosystems or human health on European scale" are taken into account (Goedkoop et al. 1996). Eco-indicator 99 (EI99) methodology was developed under the authority of the Dutch Ministry of Housing, Spatial Planning and the Environment by a number of Dutch and Swiss LCA experts and is a successor of Eco-Indicator 95. Eco-Indicator 99 is a damage-oriented method, which considers three damage categories: human health, ecosystem quality and resources (Goedkoop and Spriensma 2001).

The hierarchy of scenarios evaluated in terms of environmental impacts with the CML 2001 method is illustrated in Table 3.1. The results indicate that S1 has negative impacts when considering global warming potential (GWP), eutrophication potential (EP), photochemical ozone creation potential (POCP) and acidification potential (AP). In the S7 case, it showed positive impacts for GWP, POCP, AP and human toxicity potential (HTP), with the exception of EP for which S7 has negative impacts.

The environmental impacts of the scenarios evaluated with CML96, EDIP1997, EDIP2003, EI95, EI99 and HA methodologies are presented in Table 3.2. The results show that S1 has negative impacts according to all methodologies, while S7 has positive impacts on the environment (CML96, EI99, HA). S8 has positive impacts according to the results obtained for EDIP1997 and EDIP 2003, while EI99 indicates that scenario S6 has the greatest positive impact on the environment.

Scenario 7 can be considered the most favourable scenario from the environmental point of view, which includes **separate collection by waste fractions, recycling of materials, composting of organic waste and incineration of the residual waste**.

Increasing the separate collection of solid waste will lead to an increase in the amount of waste for recycling that ultimately leads to benefits for the environment and economical retribution. The amount of mixed waste will decrease with the increase of separately collection intensity of municipal solid waste.

3.5.2 Looking at a Comparison with the Waste Management of Twence Company in the Twente Region of the Netherlands

In contrast to Romania, the transition from the traditional waste management towards a circular economy in the Netherlands is underway (OECD 2015). According to OECD (2015), municipal waste generation in OECD countries was 520 kg/inhabitant in 2013. In 2013 there were collected 55 kg/inhabitant of paper for recycling and 75 kg/inhabitant of compostable waste (CBS 2015).

The Dutch waste management is based on Lansink's Ladder incorporated in 1994 in the Dutch legislation, entailing prevention of waste as much as possible, recovering of the valuable raw materials from any waste created, generation of energy by incinerating the residual waste and landfilling the rest of remaining waste (EVD 2008). The main actions of the Dutch policy to discourage landfilling and to improve safety were *landfill decree*, which corresponds to the description of higher technical requirements as much as standards and financial covering of post-closure costs, *landfill ban* that indicates no dumping rules for 32 types of wastes and *landfill tax and reorganisation of the landfill sector* demand that only high standards landfills were financially viable to continue their operation (de Jong 2011; SenterNovem 2006; SenterNovem 2009). To encourage the prevention, reuse and recycling, companies are required to separate waste as much as possible (as condition for their permit). On the other hand, municipalities received task levels for household separation and signed voluntary agreements with the industry on reducing packaging. More importantly, the municipalities introduced the producers' responsibility principle of all those wastes generated of their products (SenterNovem 2009). The policy of the Netherlands played an important role in the development of the professional waste market. Recycling options are cheaper than landfilling, various type of waste can be useful resources, wood can be reused or combusted, metals can be sold to the metal industry, etc. (SenterNovem 2009).

Just 2% of the all Dutch waste was landfilled in 2008, and the number of landfill sites has been reduced from 90 in 1991 to 22 in 2008 (EVD 2008; DWMA 2010a). The landfill tax was introduced in 1995 to divert MSW from landfill. In 2002 the landfill tax in the Netherlands was 79 euro/ton of waste, while in 2010 it was 107.49 euro per ton solid waste (DWMA 2010a; EEA 2013a). From January 2012 the landfill tax has been eliminated (EEA 2013a). Landfilling decreased from 2.1% in 2007 to only 0.3% in 2010 (EEA 2013a). In Enschede where the Twence Company operates, landfilling of household waste was 0% in 2008 (CBS 2011).

Material recycling in the Netherlands has been increasing in the last decades, so that in 2009 all recycling targets were easily met (DWMA 2010a, b). The recycling rates for packaging waste were 80% for plastic, 95% paper and cardboard, 88% metal, 92% glass and 38% wood (DWMA 2010b). Almost 88% of waste generated in the Netherlands in 2010 was recycled (van Eijk 2012).

In the Netherlands more than 95% of construction and demolition waste is recycled (DWMA 2010a). Waste is considered a new raw material, and recycling is a key element in the sustainable use of raw materials. This way of operating reduces the amount of natural resources used due to any increasing of quantity of waste recycled (DWMA 2010c). However, it is not considered *economically feasible to recycle everything*. According to the CBS (2011), a fraction of 20% from the total municipal waste generated in Enschede, Netherlands, was recycled in 2008.

Composting of vegetable, garden and fruit waste in the Netherlands is done for many years, so that almost 600,000 tons of compost are produced yearly by processing of 1.3 million tons of organic waste (biowaste and garden waste) (DWMA 2010d). The Twence Company from the Twente Region (which has a similar number of inhabitants as Iasi County) is able to treat 30,000 tons of green waste each year through composting process (Twence 2009). The CO_2 saving potential for traditional composting is 50 kg of CO_2 eq. (van Haeff 2007).

Another method for the treatment of organic waste used in the Netherlands, as t large, is the anaerobic digestion, as a "combination of energy generation and composting" (DWMA 2010d). The process generates energy in the form of biogas and the residue can be converted into compost. Compost resulted from anaerobic digestion has a better quality than that produced by the conventional process. The combination of energy saving and CO_2 reduction also responds to the sustainability criteria (DWMA 2010a).

Anaerobic digestion of waste began in the Netherlands in 1997 in Orgaworld Plant with a capacity of 35 ktons of waste (DWMA 2010a). An anaerobic digester with the following components, reactor, gas buffer and gas engines, was under construction at the Twence Company in 2011 with a capacity of 50,000 t/y of waste (de Jong 2011). In 2014 a total of 35,000 tonnes of organic waste was processed (Twence 2014).

In the Netherlands, 34% of the municipal solid waste was incinerated to obtain energy in 2008 (EVD 2008). The Dutch emissions from waste incineration are already below EU limit values (DWMA 2010b). In 2011 the optimal use of resources at the Twence Company results in incineration of 520,000 t of waste yearly, as well as 200,000 t of demolition wood producing of 407,000 MWhe electricity (sufficient to supply approximately 123,000 households) and 421,000 MWhe heat. All these mean a reduction of 269,000 ton/year CO_2 (de Jong 2011; EVD 2008). In 2014 a total of 733,000 tons of steam was supplied to AkzoNobel, which produces salt by evaporating brine with the aid of steam (Twence 2014). Heat coming from Twence is used by 7000 households from Enschede; for that 169,000 MWh thermal energy was supplied in 2014 (Twence 2014). The waste management in the Netherlands is practically unique in the world with a good waste processing market including recycling of impressive amounts of waste and incineration of the most of the remaining quantity of waste (EVD 2008). Therefore, the Dutch model can be used in the evaluation of the Romanian waste management system in order to shift from conventional waste technologies to a sustainable waste management system.

3.6 Opportunities for Solid Waste Management in Romania in the Context of Circular Economy

In this section we discussed about some policy potential for improvement of waste management alternatives, zero waste to landfill aspects and possibilities to implement this concept in Romania. Some examples of waste recycling campaigns performed in Romania were also given by us in this section in order to underline that efforts are being made with the purpose to involve population in the selective collection process.

3.6.1 Policy: Potential for Improvement

As discussed above, circular economy is promoting an increase in resource productivity and proposes to change the model "take-make-dispose" to "take-make-recreate" model, focusing on recycling, reuse and remanufacturing (Florin et al. 2015). When applied, these activities will reduce the footprints of materials consumed and natural resources needed to manufacture new products (Florin et al. 2015). So far there are a number of policies and measures at European, national, regional and local levels, and also a series of efforts are made by different stakeholders and interested parties for transition to a circular economy (EC 2014a).

In July 2014 *Circular Economy Package* was published. The policy discussions related with this package includes (EC 2011, 2012, 2014a; EEA 2013b):

- Revised legislative proposals on waste
- Implementation of the *Roadmap to a Resource Efficient Europe* (7th Environmental Action Programme)
- Advancing the green *economy within and beyond the EU*
- Production of renewable biological resources and their conversion into vital products and bioenergy (*Bioeconomy Strategy*)
- Implementation of the *Europe 2020 Strategy*

These policies can serve as a good base, but they are insufficient for progress of circular economy in the EU, which requires an integrated approach (EC 2014a). Better implementation and enforcement of existing regulation (waste framework directive, packaging waste directive, landfill directive, etc.) can facilitate the transition of circular economy and also would reduce the differences in performance.

Revisions of current regulations and new regulation (new targets, restrictions or bans, etc.) are necessary for better support of circular economy (EC 2014a).

EU waste policy aims: "to ensure that by 2020 waste is managed as a resource, to achieve an absolute decline of waste generated per capita, to ensure high quality recycling, to limit energy recovery to non-recyclable materials, virtually eliminate landfilling and eradicate illegal shipments of waste" (EC 2011; EEA 2013b).

However, there are considerable differences between member states in terms of performance (EC 2014b). For example, in the Netherlands, the targets for material recovery address an increase in total waste recovery rate from 77% to more than 83% between in 2000 and 2012, which was achieved in 2005, and increasing the total waste recovery rate from 83% to 85% between 2006 and 2015 also already accomplished by 2010 (OECD 2015). Based on the fact that the recovery rates are so high, the intention is to focus in the next years on increasing the quality of recycled materials.

In comparison with the EU average, the waste sector in Romania is very far behind. In 2011 the recycling rate in Romania was 7% of the total collected waste (Cioca et al. 2015). The National Recycling Materials Commission in 2007 approved 185 recovering companies for waste of paper and cardboard, 266 recovering companies for plastics waste and 55 recovering companies for glass (BALKWASTE 2010). If collected and sorted materials will eventually be used in specific branches of industry, the recovery and recycling will be successful, and the Romanian government is aware of this (NWMP 2004).

3.6.2 Zero Waste to Landfill

Zero waste is considered an alternative solution for waste management problems. This concept can stimulate sustainable production and consumption, recovery and recycling and restricts incineration and landfilling (Zaman 2015). In the negotiations on the circular economy is mentioned that "a way to close the loop is to ban landfilling" (ZWE 2015). It was demonstrated that the European countries (Denmark, Germany, the Netherlands, Norway, Sweden, Austria and Switzerland), which applied this measure, predominantly increased the waste incineration rate, technology that also generates waste (fly and bottom ashes) (ZWE 2015). According to ZWE (2015), the concept of "zero waste to landfill" or landfill ban contributes to a perfect linear economy and considered that "high tax on landfill and waste to energy incineration combined with a lower tax on the landfilling of stabilised waste is more effective in diverting waste towards prevention, preparation for re-use and recycling than a landfill ban".

Under this conceptual framework, the question is *if it would be possible to implement a zero waste to landfill system in Romania.*

As prior mentioned, landfilling is the common method for municipal solid waste disposal in Romania, since in urban areas there is at least one waste landfill (BALKWASTE 2010; Ghinea 2012). More than 95% of the collected wastes were disposed in landfills in 2009, while all rural dumpsites were closed in the same year. On the other hand, a number of 29 compliant landfills were built in 2010 (ENVIROPLAN S.A. 2012). According to EC (2013), there were 80 non-compliant landfills officially operating in 2013, which must be closed before July 2017. The European Commission gave the following recommendations to Romania (EC 2013): introduction of a landfill tax and increasing this tax in order to divert waste from landfills; expanding and improving the cost-effectiveness, monitoring and transparency of existent schemes; implementation of a biowaste strategy in order to divert biodegradable waste from landfills; compliance with legal requirements; development of separate collection infrastructure and schemes; comprehensive awareness raising campaigns on separate collection and proper waste management; and improvement of allocation of available EU funding in order to support waste prevention, reutilisation and recycling.

Romania began with very small steps to apply a landfill tax from the 1st of January 2016: the landfill tax in Romania is 80 RON (€ 18) per ton, and it will increase to 120 RON/ton (€ 27) in 2018 (Econet Romania 2016). The main obstacles in introducing zero waste in Romania are (Tartiu and Petrache 2009) unwarrantable infrastructure for the waste management, low involvement of citizens and economic agents in sustainable waste management, the complexity of zero waste system implementation and lack of experience.

We believe that the involvement of authorities, stakeholders and the population as at large by being one of the key actors in the solid waste management system will lead to changes in the Romanian system. The action should start from the population, which must separate their waste at home in waste fractions. Another measure which can be taken is alternating the municipal collection of organic waste in 1 week and that of other wastes in the next week. Also, the collection of paper waste can be performed by different associations.

Local authorities and sanitation companies should facilitate solid waste collection on waste fractions to recycling companies which can deal with different type of waste and support recycling of materials. The authorities at country level should deal with legal requirements, taxes and funds to support the waste management system. It may take a long time but with the involvement of all these actors may ultimately help in reaching the *zero waste* purpose. The population should become aware about the environmental impact of waste generated and that solid waste can be used as resources.

3.6.3 Public Perception, Awareness and Participation

In recent years various Romanian companies have developed waste recycling campaigns to involve the population in the process of selective collection and to inform them about the importance of selective waste collection. Some examples are given below.

In 2014 Carrefour Romania and Green Group (the largest investor in the green industry in Romania, bringing together five companies specialised in management, recycling and recovery of waste) launched in Buzau, Romania, the pilot project of smart waste collection SIGUREC 1 for collecting plastic bottles, aluminium cans, glass bottles, bottles PE, PP, Tetra Pak, packaging foils, paper, cardboard and waste electrical and electronic equipment. They also tested the solution for collecting PET bottles and aluminium cans in Carrefour Baneasa (Bucharest) and Carrefour Braila using special machines placed inside shopping centres. SIGUREC stations collected over 1000 tons wastes from 1 August to 31 December 2014, which were sent directly to recycling factories in Romania. The quantities of waste taken through the SIGUREC in the last 5 months of 2014 were 772,063 units of PET packaging, cans and glass, 310,121 kg of waste electric and electronic equipment and 259,063 kg of paper/cardboard (Carrefour 2015).

The main advantage of the project consists in that the general public has realised that waste is carrying value, since people received value vouchers that can be used for shopping in Carrefour: 0.05 RON/PET, 0.03 RON/dose aluminium, 0.01 RON/glass bottle and a variable amount per kg for WEEE, depending on the device type taught (e.g. refrigerator 0.8 RON/kg, iron 0.6 RON/kg, laptop 1.5 RON/kg) (Carrefour 2015). In October 2015 the same project was also inaugurated in Iasi.

Another project "Green for recycling!" for information and awareness on the separate collection of packaging waste was launched by Eco-Rom packaging, in partnership with the Intercommunity Development for Sanitation A.D.I.S Iasi during July 25 to August 15, 2015. Under the slogan "Iasi collects separately packaging waste!", people from Iasi were asked to separately collect packaging waste. People who have separately collected packaging waste properly were awarded with frisbee, bracelets, sunglasses and many other prizes. Also, campaigns include radio spots and competitions, in which people from Iasi learned useful information about the separate collection and recycling of packaging waste (Eco-Rom packaging 2016).

"ComPETition for a better future" was also a campaign that the Lidl company carried out all over the country during June–October 2015. Population had to bring PET packaging during the campaign at recycling points located in Lidl shops parking and received awards consisting of in a LIDL shopping voucher and others (Lidl 2015).

On 10th of March 2016 in Iasi, the first centre of municipal waste collection was opened. The inauguration of the centre was accompanied by an extensive information and awareness campaign in the months thereafter. By the 31st of March, citizens who brought recyclable waste to the centre won prizes and were entered in a raffle with ten attractive prizes (Ecotic 2016).

Even if these campaigns are publicised and the population participates to win those awards, there is still much to do to increase awareness. For example, after applying a questionnaire among people in Suceava city in the NE region of the country (100 people surveyed: 57% females and 43% males represented by young people up to 30 years), only 84% of females interviewed consider that there is a problem with waste disposal, while only 69% of males agree on this. Sixty-five percent of females responded that recycling is very important, while 69% of male respondents believe this (Ghinea et al. 2016b). The respondents suggested that installation of recycling bins for different types of waste in every residential area increases the recycling habit of people. Also, the awareness level and education plays a very important role in solid waste recycling participation as reported by Ghinea et al. in 2016b.

3.7 Conclusions

The idea of the circular economy in Europe is achieving further consideration and recognition. The key actions which can support the circular economy are the prevention and decrease of waste generation rate and the intensification in the reuse and recycling of products, but also zero waste. However, there are considerable differences between member states in terms of performance.

Although there are challenging targets in the view of waste minimisation, based on long-term approaches addressing waste management and recycling, Romania is far behind in terms of implementation of the circular economy principles. As demonstrated by the waste management analysis in Iasi, landfilling continues to be the prevalent management practice, although relevant improvements were made after Romania's admission as an EU member state. The most favourable waste management scenario out of eight LCA assessments (environmental point of view) is one which includes separate collection of waste fractions, recycling of materials, composting of organic waste and incineration of the residual waste.

Increasing the separate collection of solid waste will lead to an increase in the amount of waste for recycling that ultimately leads to benefits for the environment. The amount of mixed waste will decrease with the increase of separate collection intensity of municipal solid waste. If collected and sorted, materials will eventually be used in specific branches of industry, the recovery and recycling will be successful and Romania is aware of this. The involvement of authorities, stakeholders and last but not least of the population is a key factor to lead to changes in the Romanian solid waste management system. The practices leading towards zero waste to landfill can be integrated to the existing waste management strategies with the aim to generate additional economic value to Romanian users and operators of SWM system.

Acknowledgement This work was elaborated with the support of EURODOC "Doctoral Scholarships for Research Performance at European Level" project ID_59410, financed by the European Social Found and Romanian Government, GaBi4: Software and database for Life Cycle Engineering, PE INTERNATIONAL GmbH and with the support of two Grants of the Romanian National Authority for Scientific Research, CNCS-UEFISCDI: project PN-III-P4-ID-PCE-2016-0683, contract 65/2017, and project PN-III-P2-2.1-BG-2016-0130, contract 64BG/30.09.2016.

References

AG (2009) Living sustainably: the Australian Government's National Action Plan for Education for Sustainability. Australian Government. Department of the Environment, Water, Heritage and the Arts

BALKWASTE (2010) Action 3: evaluation of waste infrastructure in Region 1 North-East, Romania. Establishment of Waste Network for sustainable solid waste management planning and promotion of integrated decision tools in the Balkan Region (BALKWASTE), Bacau. http://www.balkwaste.eu/?page_id=21. Accessed 14 Nov 2011

Bayer S, Méry J (2009) Sustainability gaps in municipal solid waste management: a case study for landfills. Environ Dev Sustain 11:43–69. https://doi.org/10.1007/s10668-007-9097-0

Best A, Blobel D, Cavalieri S, Giljum S, Hammer M, Lutter S, Simmons C, Lewis R (2008) Potential of the ecological footprint for monitoring environmental impacts from natural resource use. Report to the European Commission, DG Environment. Final Report, May 2008, EC Commission, Brussels

Björklund A, Finnveden G (2005) Recycling revisited-life cycle comparisons of global warming impact and total energy use of waste management strategies. Resour Conserv Recycl 44:309–317. https://doi.org/10.1016/j.resconrec.2004.12.002

Carrefour (2015) Carrefour Romania and Green Group announces results of selective waste collection system SIGUREC, (in Romanian). https://www.carrefour.ro/despre-noi/stiri-presa/noutati/carrefour-romania-si-green-group-anunta-rezultatele-sistemului-de-colectare-selectiva-a-deseurilor-sigurec/. Accessed 10 Mar 2016

CBS (2011) Statistics Netherlands. Centraal Bureau voor de Statistiek. http://www.cbs.nl/en-GB/menu/cijfers/default.htm. Accessed 12 Apr 2011

CBS (2015) Trends in the Netherlands 2015. Statistics Netherlands

CEPI (2012) Paper Recycling, Monitoring Report, 2012. Confederation of European Paper Industries

Chang N-B, Pires A, Martinho G (2011) Empowering systems analysis for solid waste management: challenges, trends, and perspectives. Crit Rev Environ Sci Technol 41:1449–1530

Cioca LL, Ivascu L, Rada EC, Torretta V, Ionescu G (2015) Sustainable development and technological impact on CO2 reducing conditions in Romania. Sustainability 7:1637–1650. https://doi.org/10.3390/su7021637

COM (2011) Roadmap to a resource efficient Europe. Communication from the Commission to the European Parliament, the Council, the European Economic and Social Committee and the Committee of the Regions, Brussels. http://eur-lex.europa.eu/legal-content/EN/TXT/PDF/?uri=CELEX:52011DC0571&from=EN. Accessed 14 Mar 2016

COM (2014) Toward a circular economy: a zero waste Programme for Europe. Communication from the Commission to the European Parliament the Council, the European Economic and Social Committee and the Committee of the Regions. http://eur-lex.europa.eu/resource.html?uri=cellar:aa88c66d-4553-11e4-a0cb-01aa75ed71a1.0022.03/DOC_1&format=PDF. Accessed 14 Mar 2016

COM (2015) Closing the loop – an EU action plan for the Circular Economy. Communication from the Commission to the European Parliament. The Council, The European Economic and Social Committee and the Committee of the Regions, European Commission. Brussels

Comanita ED, Ghinea C, Rosca M, Simion IM, Petraru M, Gavrilescu M (2015) Environmental Impacts of Polyvinyl Chloride (PVC) Production Process. E-Health and Bioengineering Conference (EHB), 2015, 19–21 November 2015, pp 1–4. https://doi.org/10.1109/EHB.2015.7391486

Comanita ED, Hlihor MR, Ghinea C, Gavrilescu M (2016) Occurrence of plastic waste in the environment: ecological and health risks. Environ Eng Manag J 15:675–685

Davidescu A, Strat VA (2014) An empirical analysis of regional poles from the perspective of Romanian sustainable development. Proced Econ Finan 10:114–124. https://doi.org/10.1016/S2212-5671(14)00284-6

de Jong W (2011) Twence from waste to energy. Visit Romanian PhD student from Twente University presentation. Twence Company, Hengelo
DEFRA (2011) Applying the waste hierarchy: evidence summary. Environment Agency – Department for Environment Food and rural Affairs, London
Doba A, Nistorescu M, Pastravanu L, Petruc I, Simionescu I (2008) Strategic plan for integrated waste management in Iasi County, (in Romanian). http://www.scribd.com/doc/13769639/Planul-Judetean-de-Gestionare-al-Deseurilor. Accessed 10 Jan 2010
DWMA (2010a) MEP's about European Agenda. Avalforum. no. 1. Dutch Waste Management Association (DWMA)
DWMA (2010b) Waste companies and producers cooperate on sustainable design. Avalforum. no. 4. Dutch Waste Management Association (DWMA)
DWMA (2010c) Sustainable use of materials on the EU Agenda. Avalforum. no. 3. Dutch Waste Management Association (DWMA)
DWMA (2010d) High time for a European Biowaste Directive. Avalforum. no. 2. Dutch Waste Management Association (DWMA)
EC (1994) European Parliament and Council Directive 94/62/EC of 20 December 1994 on packaging and packaging waste. Off J Eur Commun 365:10–23
EC (2000) Directive 2000/76/EC of the European parliament and of the council of 4 December 2000 on the incineration of waste. Off J Eur Commun L, 332/91
EC (2005) EU waste policy. The story behind the strategy. European Commission. http://ec.europa.eu/environment/waste/strategy.htm. Accessed 17 May 2012
EC (2011a) Report from the Commission to the European Parliament, the Council, the European Economic and Social Committee and the Committee of the Regions on the Thematic Strategy on the prevention and recycling waste, COM (2011), 13 final, Brussels
EC (2011b) Roadmap to a resource efficient Europe. European Commission, Brussels
EC (2012) Innovating for a bioeconomy for Europe sustainable growth. European Commission, Brussels
EC (2013) Support to member states in improving waste management based on assessment of member states' performance. European Commission, Brussels
EC (2014a) Scoping study to identify potential circular economy actions, priority sectors, material flows and value chains. European Commission, Luxembourg
EC (2014b) Towards a circular economy: a zero waste programme for Europe. European Commission, Brussels
EC (2016) Review of the Thematic Strategy on the prevention and recycling of waste. http://ec.europe.eu/environment/waste/strategy.htm. Accessed 14 Mar 2016
EC Council Directive (1999) Council Directive 1999/31/EC of 26 April 1999 on the landfill of waste. Off J Eur Commun L, 182/1, 16.7.1999, Brussels
EC Council Directive (2002) Directive 2002/96/EC of the European Parliament and of the Council of 27 January 2003 on waste electrical and electronic equipment (WEEE). Off J Eur Commun L, 037, 13/02/2003 P. 0024 – 0039
EC Council Directive (2008), Directive 2008/98/EC of the European Parliament and of the Council of 19 November 2008 on waste and repealing certain Directives. The Official Journal of the European Communities L, 312/3, 22.11.2008, Brussels
Econet Romania (2016) Landfill tax in Romania. http://www.econet-romania.com/en/single-news/731/starting-with-2016-the-landfill-tax-in-romania-will-raise-to-80-lei-18-per-ton.html. Accessed 10 Mar 2016
Eco-Rom packaging (2016) Green for recycling in Iasi! (in Romanian). http://ecoromambalaje.ro/verde-la-reciclare-in-iasi/. Accessed 10 Mar 2016
Ecotic (2016) Premiere for Romania: in Iasi was opened the first center Municipal Waste Collection, (in Romanian). http://www.ecotic.ro/stiri/premiera-pentru-romania-la-iasi-s-a-deschis-primul-centru-municipal-de-colectare-a-deseurilor/. Accessed 10 Mar 2016
EEA (2013a) Municipal waste management in the Netherlands. European Environment Agency (EEA)

EEA (2013b) Towards a green economy in Europe, EU environmental policy targets and objectives 2010–2050. European Environmental Agency
EEC (1993) Council Regulation (EEC) No. 259/93 on the supervision and control of shipments of waste within, into and out of the European Community. Off J Eur Commun L, 30, 6 February 1993, pp. 1–28
ENVIROPLAN S.A. (2012) Romania – Identification of future waste management projects (2014–2020)-Task 1 Baseline report
EPAIS (2014a) Environmental Status Report for 2014 in Iasi County. Environmental Protection Agency Iasi (EPAIasi), (in Romanian). http://apmis.anpm.ro/ro/rapoarte-anuale1. Accessed 10 Oct 2015
EPAIS (2014b) Local action plan for the environment Iasi County 2014–2018. Environmental Protection Agency Iasi (EPAIasi), (in Romanian). http://www.anpm.ro/documents/839616/2172756/PLAM+2014-2018-+judetul+Iasi.pdf/0c22eb6a-819e-4fe8-9b38-7e38a32bf86a. Accessed 10 Oct 2015
ERPC (2011) European declaration on paper recycling 2011–2015. European Recovered Paper Council. http://www.paperforrecycling.eu/uploads/Modules/Publications/DeclarationonPaperRecycling_WebVersion.pdf. Accessed 31 Mar 2016
ERPC (2015). Paper recycling monitoring report. European Recovered Paper Council. http://www.paperrecovery.org/uploads/Modules/Publications/Final_MonitoringReport2014.pdf. Accessed 31 Mar 2016
Eurostat (2010) Waste statistics. http://epp.eurostat.ec.europa.eu/statistics_explained/index.php/Waste_statistic. Accessed 10 Oct 2015
Eurostat (2015) Population and population change statistics. Eurostat Statistics Explained. http://ec.europa.eu/eurostat/statistics-explained/index.php/Population_and_population_change_statistics. Accessed 31 Mar 2016
EVD (2008) Made in Holland. Waste a valuable resource. Agency for International Business and Cooperation. http://www.hollandtrade.com/publications/made-in-holland/mih.asp?bron=waste. Accessed 12 Apr 2011
FERVER (2015) Glass recycling: years of improvement. European Federation of Glass Recyclers. http://www.glassmanevents.com/europe/content-images/misc/FERVER.pdf. Accessed 31 Mar 2016
FEVE (2015) European container glass industry welcomes long awaited circular economy package. http://www.feve.org/images/stories/2015/feve-paper-circ-econ-pack-question-2.pdf. Accessed 31 Mar 2016
Florin N, Madden B, Sharpe S, Benn S, Agarwal R, Perey R, Giurco D (2015) Shifting business models for a circular economy: metals management for multi-product-use cycles, UTS, Sydney. http://wealthfromwaste.net/wp-content/uploads/2015/11/P3-FINAL-SHIFTING-BUSINESS-MODELS-FOR-CE-ONLINE.pdf. Accessed 31 Mar 2016
Frischknecht R, Jungbluth N, Althaus H-J, Bauer C, Doka G, Dones R, Hischier R, Hellweg S, Humbert S, Köllner T, Loerincik Y, Margni M, Nemecek T (2007) Implementation of life cycle impact assessment methods. Ecoinvent report no. 3, v2.0. Swiss Centre for Life Cycle Inventories, Dübendorf
Gavrilescu M (2011) Sustainable industrial production. Ecozone Press, Iasi
GD (2013) Government Decision no. 870/2013 approving the National Strategy for Waste Management 2014–2020. Official Gazette of Romania no. 750 of December 4, 2013
GEO (2005) Government Emergency Ordinance 195/2005 on environmental protection. The Government of Romania, Bucharest
Gharfalkar M, Court R, Campbell C, Ali Z, Hillier G (2015) Analysis of waste hierarchy in the European WASTE DIRECTIVE 2008/98/EC. Waste Manag 39:305–313. https://doi.org/10.1016/j.wasman.2015.02.007
Ghinea C (2012) Waste management models and their application to sustainable management of recyclable waste. PhD Thesis, Gheorghe Asachi Technical University of Iasi, Romania

Ghinea C, Gavrilescu M (2010a) Decision support models for solid waste management – an overview. Environ Eng Manag J 9:869–880
Ghinea C, Gavrilescu M (2010b) Models for sustainable waste management. Bulletin of the Polytechnic Institute of Iasi, Chemistry and Chemical Engineering Section, LVI (LX) (2):21–36
Ghinea C, Petraru M, Bressers HTA, Gavrilescu M (2012) Environmental evaluation of waste management scenarios – significance of the boundaries. J Environ Eng Landsc Manag 20:76–85. https://doi.org/10.3846/16486897.2011.644665
Ghinea C, Simion IM, Gavrilescu M (2013) Solid waste recycling for remanufacturing and bioremediation. Bulletin of the Polytechnic Institute of Iasi, Chemistry and Chemical Engineering Section, LIX(2), 35–48
Ghinea C, Petraru M, Simion IM, Sobariu D, Bressers HTA, Gavrilescu M (2014) Life cycle assessment of waste management and recycled systems. Environ Eng Manag J 13:2073–2085
Ghinea C, Dragoi EN, Comanita E-D, Gavrilescu M, Campean T, Curteanu S, Gavrilescu M (2016a) Forecasting municipal solid waste generation using prognostic tools and regression analysis. J Environ Manag 182:80–93. https://doi.org/10.1016/j.jenvman.2016.07.026
Ghinea C, Comanita DE, Gavrilescu M (2016b) Population perception on reduce, reuse and recycling of household solid waste. In: 11th international conference on "Environmental legislation, Safety Engineering and Disaster Management" ELSEDIMA, 26–28 May 2016, Cluj-Napoca, Romania
Ghisellini P, Cialani C, Ulgiati S (2015) A review on circular economy: the expected transition to a balanced interplay of environmental and economic systems. J Clean Prod 114:11–32. https://doi.org/10.1016/j.jclepro.2015.09.007
Goedkoop M, Spriensma R (2001) The Eco-indicator 99 a damage oriented method for life cycle impact assessment. Methodology Report. Pre Consultants b.v., Amersfoort
Goedkoop M, Demmers M, Collignon M (1996) Eco-indicator 95. Manual for designers, The Netherlands
Haas W, Krausmann F, Wiedenhofer D, Heinz M (2015) How circular is the global economy? an assessment of material flows, waste production, and recycling in the European Union and the World in 2005. J Ind Ecol 15:765–777. https://doi.org/10.1111/jiec.12244
Hall D (2010) Waste Management in Europe: framework, trends and issues. PSIRU, business school. University of Greenwich, Park Row
Hauschild M, Potting J (2005) Spatial differentiation in life cycle impact assessment – the EDIP2003 methodology. Environmental news no. 80
Holmberg J, Lundqvist U, Robert K-H, Wackernagel M (1999) The ecological footprint from a system perspective of sustainability. Int J Sustain Dev World Ecol 6:17–33. https://doi.org/10.1080/13504509.1999.9728469
Iasi County Council (2009) Long-term investment plan for the period 2008–2038 on integrated waste management. http://www.icc.ro/activitate/mediu/MasterP/MasterPlan.pdf. Accessed 14 Nov 2011
INS (2011) County statistics. Iasi County Statistics. http://www.iasi.insse.ro/main.php?id=373. Accessed 14 Nov 2011
INS (2015) County statistics. Iasi County Statistics. http://www.iasi.insse.ro. Accessed 10 Oct 2015
INS (2016) County statistics. Iasi County Statistics. http://www.iasi.insse.ro. Accessed 31 Mar 2016
ISWA (2015) Circular economy: trends and emerging ideas. International Solid Waste Association, Vienna
Jávor B (2015) Circular economy – what we can achieve with ambitious recycling targets and what is crucially needed beyond. In: Circular economy in Europe towards a new economic model, The European Files, September 2015
JRC (2011) Supporting environmentally sound decisions for bio-waste management. A practical guide to life cycle thinking (LCT) and life cycle assessment (LCA). European Commission, Joint Research Centre, Institute for Environment and Sustainability, Ispra
JRC European Commission (2010) Analysing of existing environmental impact assessment methodologies for use in life cycle assessment. Joint Research Centre Institute for Environment and Sustainability, ILCD handbook

LAP 2 (2008) National Waste Management Plan (LAP) 2009–2021. Ministry for Housing, Spatial Planning and the Environment (VROM), Hague

Law 211 (2011) Law No. 211 of 15 November 2011 on waste regime. The Romanian Parliament, published in Official Gazette no. 837 of 25 November 2011

Lidl (2015) Competition Lidl, (in Romanian). https://www.competitialidl.ro/Regulament_ComPETitia.pdf. Accessed 10 Mar 2016

Luca FA, Ioan CA (2014) Implementation of green marketing in the analysis of municipal waste produced in Romania, correlated with environmental policy management. Environ Eng Manag J 13:3131–3142

Mac Arthur Foundation (2012) Towards a circular economy: business rationale for an accelerated transition. Ellen Mac Arthur Foundation Corves, UK. https://www.ellenmacarthurfoundation.org/assets/downloads/TCE_Ellen-MacArthur-Foundation_9-Dec-2015.pdf. Accessed 10 Mar 2016

Maguin D (2015) Why Europe needs a market-driven circular economy? In: Circular economy in Europe towards a new economic model, The European Files, September 2015

MECC (2016) Waste management, Romanian Ministry of Environment and Climate Change. http://www.mmediu.ro/beta/domenii/gestionarea-deseurilor/. Accessed 10 Mar 2016

NSDSR (2008) National Sustainable Development Strategy Romania 2013–2020–2030. Government of Romania – Ministry of Environment and Sustainable Development. United Nations Development Program – National Centre for Sustainable Development, Bucharest

NWMP (2004) National Waste Management Plan. Ministry of Environment and Water Management, Bucharest

NWMS (2004) National Waste Management Strategy. Ministry of Environment and Water Management, Bucharest

NWMS (2013) National Waste Management Strategy. Ministry of Environment and Climate Change, Bucharest

OECD (2007) Outcome of the first OECD workshop on sustainable materials management. Working Group on Waste Prevention and Recycling, Organisation for Economic Co-operation and Development (OECD)

OECD (2010) OECD-FAO agricultural outlook 2010–2019. Organisation for Economic Co-operation and Development/Food and Agriculture Organization of the United Nations, Paris

OECD (2011a) Policy principles for sustainable materials management. Working Group on Waste Prevention and Recycling, Organisation for Economic Co-operation and Development (OECD)

OECD (2011b) Policy instruments for sustainable materials management. Working Group on Waste Prevention and Recycling, Organisation for Economic Co-operation and Development (OECD)

OECD (2015) OECD environmental performance reviews The Netherlands 2015

Oroian I, Odagiu A, Braşovean I, Burduhos P (2009) Waste management in Romania. Case study: implementing National Strategy on waste management. ProEnvironment 2:235–242

Papachristou E, Hadjianghelou H, Darakas E, Alivanis K, Belou A, Ioannidou D, Paraskevopoulou E, Poulios K, Koukourikou A, Kosmidou N, Sortiko K (2009) Perspectives for integrated municipal solid waste management in Thessaloniki, Greece. Waste Manag 29:1158–1162. https://doi.org/10.1016/j.wasman.2008.04.015

Petrescu V, Ciudin R, Isarie C, Nederita V (2010) Environmental engineering education – waste management in Romania. World Trans Eng Technol Edu 8:107–110

Pires A, Martinho G, Chang NB (2011) Solid waste management in European countries: a review of systems analysis techniques. J Environ Manag 92:1033–1050. https://doi.org/10.1016/j.jenvman.2010.11.024

PlasticsEurope (2015a) Towards a life cycle driven circular economy. http://www.plasticseurope.org/documents/document/20150617095930-plasticseurope_view_paper_circular_economy_june_2015_final.pdf. Accessed 31 Mar 2016

PlasticsEurope (2015b) Plastics – the Facts 2015 An analysis of European plastics production, demand and waste data. http://www.plasticseurope.org/documents/document/20150227150049-final_plastics_the_facts_2014_2015_260215.pdf. Accessed 31 Mar 2016

Rossy A, Jones PT, Geysen D, Bienge K (2010) Sustainable materials management for Europe, from efficiency to effectiveness. Report for the informal Environmental Council on the 12th and 13th of July 2010, Brussels
RWMP – NE (2006) Regional waste management plan, region 1 NE. Ministry of Environment and Water Management, Bucharest
Schiopu AM, Apostol I, Hodoreanu M, Gavrilescu M (2007) Solid waste in Romania: management, treatment and pollution prevention practices. Environ Eng Manag J 5:451–465
SenterNovem (2006) The Dutch waste profile 1990–2005. http://www.senternovem.nl. Accessed 12 Apr 2011
SenterNovem (2009) Progress in Dutch waste management: 1990–2005. Twinning Domestic Waste Management Project. http://www.twinning-waste-bacau.ro. Accessed 12 Apr 2011
SEPA (2008) Swedens environmental objectives in brief. Swedish Environmental Protection Agency, Sweden
Singh J, Ordonez I (2016) Resource recovery from post-consumer waste: important lessons for the upcoming circular economy. J Clean Prod 134:342–353. https://doi.org/10.1016/j.jclepro.2015.12.020
SME (2003) EU priorities to achieve the Swedish environmental quality objectives. Swedish Ministry of the Environment, Government Communication 2003/04:9
Tartiu EV, Petrache AM (2009) The zero waste system – an enduring solution for the waste management in Romania? http://www.iswa.org/uploads/tx_iswaknowledgebase/wm_2009_1-300.pdf. Accessed 31 Mar 2016
Twence (2009) Sustainability report. http://www.twence.nl/en/over%20twence/Download_Annual_Reports.doc/. Accessed 12 Apr 2011
Twence (2014) Forging links. Public annual report
US EPA (2007) Municipal solid waste generation, recycling, and disposal in the United States: facts and figures for 2007. http://www.epa.gov/waste/nonhaz/municipal/pubs/msw07-fs.pdf. Accessed 12 Apr 2011
van Eijk F (2012) Sustainable waste management. Vice-chair DWMA, Rotterdam, 24 October 2012
van Haeff J (2007) Digestion and composting in the Netherlands. Attero. http://www.marm.es/es/calidad-y-evaluacion-ambiental/temas/prevencion-y-gestion-de-residuos/2_6_John_van_Haeff-association_Netherlands_tcm7-1688.pdf. Accessed 12 Apr 2011
Veeken A, Huisman H, van Schaik R, Suurmeijer J, Stoyanov M, Zhivkova S (2011) Handbook for biowaste management in Bulgaria. Draft. Ministry of Housing Spatial Planning and the Environment, Utrecht
von Braun J (2008) Food and financial crises: implications for agriculture and the poor. International Food Policy Resources Institute, Washington, DC
Wante, J. (2010). A European legal framework for enhanced waste management. Paper presented at 1st international symposium on enhanced landfill mining, Houthalen-Helchteren, 4-6/10/2010
WEF (2014) Towards the circular economy: accelerating the scale-up across global supply chains. World Economic Forum, Geneva
Wenzel H, Hauschild M, Alting L (1997) Environmental assessment of products: methodology, tools and case study in product development, vol I. Kluwer Academic Publisher, London
Zaman AU (2015) A comprehensive review of the development of zero waste management: lessons learned and guidelines. J Clean Prod 91:12–25. https://doi.org/10.1016/j.jclepro.2014.12.013
Zils M (2014) Moving toward a circular economy. http://www.mckinsey.com/business-functions/sustainability-and-resource-productivity/our-insights/moving-toward-a-circular-economy. Accessed 31 Mar 2016
ZWE (2015) Zero waste to landfill and/or landfill bans: false paths to a circular economy. ZeroWaste Europe. http://www.zerowasteeurope.eu/2015/11/press-release-landfill-ban-a-false-path-to-a-circular-economy/. Accessed 31 Mar 2016
ZWE (2016) Closing the loop of materials, phasing out toxics & emissions. Zero waste Europe. https://www.zerowasteeurope.eu/2016/03/press-release-eurostat-data-for-2014-confirms-need-for-european-residual-waste-target/. Accessed 2 Apr 2016

Chapter 4
Feasibility Analysis of a Cap-and-Trade System in Mexico and Implications to Circular Economy

José-Luis Cruz-Pastrana and María-Laura Franco-García

Abstract Market-based instruments, such as the cap-and-trade, have been widely used to address the increase of greenhouse gases (GHG). In line with other geographic regions, Mexico has seen the need to expand the options of market instruments to mitigate the effects of climate change. Simultaneously, there are important reasons to move towards a circular economy model. In this context, this research seeks to answer if it's feasible to implement a cap-and-trade system in Mexico as part of its climate policy. That said, firstly, it analyses and assesses the cap-and-trade system in Mexico based upon its contextual environment, its potential of implementation and its economic and environmental benefits and costs. Secondly, it highlights the implications of considering circular economy models into a cap-and-trade instrument. Through the use of marginal abatement cost curves (MACC), it was found that the percentage of measures to reduce GHG with negative cost agglomerates 57% of all the measures, which could translate into a benefit for the economy as a whole by almost 1% of GDP by 2020. As part of the conclusions, we argued that MACC results prove partially that the cap-and-trade system is a feasible option to apply in Mexico. In addition, a cap-and-trade mechanism should show a strong carbon price signal felt by end users and therefore motivates the application of circular economy principles, which are related to the introduction of innovations to enable the closing of current materials and energy loops along the supply production chain.

Keywords Cap-and-trade · Marginal abatement cost curves · Circular economy · Climate change · Climate policy

---

J.-L. Cruz-Pastrana (✉)
Instituto Tecnológico de Estudios Superiores Monterrey,
Cd López Mateos, Estado de México, Mexico

M.-L. Franco-García
University of Twente, Enschede, The Netherlands
e-mail: m.l.francogarcia@utwente.nl

M.-L. Franco-García et al. (eds.), *Towards Zero Waste*, Greening of Industry Networks Studies 6, https://doi.org/10.1007/978-3-319-92931-6_4

4.1 Introduction

In recent years, climate change has positioned as the greatest environmental thread that not only the human kind faces but the different ecosystems that inhabit the planet. The impact of climate hazards is no longer denied, and different global actions have emerged to curb greenhouse gases (GHG).

Market-based instruments have been the most penetrating approach to tackle this issue. The cap-and-trade system is based on the market to internalise the external costs of fossil energy use. This scheme has been adopted by some countries and economic blocs (such as the European Union) as a key policy in mitigating GHG emissions. While it is argued that the implementation only belongs to Annex I countries for its international commitments, proposals and pilot schemes in countries outside Annex I are beginning to emerge, such as the recent approval of the Chinese national cap-and-trade in September 2015. Mexico has been no exception, and it's currently building the institutional and legal framework conditions necessary to establish the instruments to mitigate and adapt to the effects of climate change and move towards a sustainable economy.

The content of this paper starts with a conceptual and theoretical framework of the cap-and-trade system as a market-based instrument to mitigate GHG emissions where the objectives and its operation are addressed. Later on, the design principles that should be considered when developing the scheme as a public policy are presented. After that, the feasibility of the instrument in Mexico is analysed, and the results obtained from the assessment of Mexico's marginal abatement cost curve are presented. Finally, some of the convergences, divergences and implications among market-based instruments and circular economy thinking are discussed.

4.2 Overview of the Cap-and-Trade System

Due to the lack of defined property rights over common goods, a pure market approach is practically impossible to implement in the context of climate change (Wagner 2013). That said, the cap-and-trade system was created to be a practical alternative that combines the interaction between governments (which represent society) and pollutants through a market.

In recent years, the cap-and-trade system has become not only the backbone of the climate policy in some countries but also has been the mitigation instrument that has received the biggest boost in international agendas. Betsill and Hoffmann (2011) consider that the reason of such boost relies on the fact that currently most of the discussions have moved from the design phase to a pilot or deployment programme. Quoting Simmons and Elkins (2004, 173 p.):

> As growing numbers of important actors articulate theories and implement practises that reflect a normative consensus, the legitimacy of these ideas gathers steam.

Lascoumes and Le Gales (2007) note that from the point of view of economists, cap-and-trade is interpreted as a general equilibrium model in which the feasibility of the scheme is given by their economic and environmental effectiveness. However, this "functionalist orientation", as they denote, ignores the social, political and ethical standards, which are critical for a sound market development. Moreover, recent events have shown how fragile the cap-and-trade system actually is. This occurs primarily in the context of financial crisis, increase of the scepticism about climate science and an increasingly polarised decision-making in key countries like the USA, Canada and Australia.

In a cap-and-trade system, the authorities set a priori the acceptable level of emissions; this amount is called cap and attempts to replicate the optimal level of contamination, between the social damage and the abatement cost for the polluters. Usually, the level of pollution will be given by the GHG emissions generated by the agents subject to this system.

Once the cap is set, the next step is to generate certificates, which represent the right to emit 1 tonne of GHG. Emission certificates are allocated to pollutants either free of charge (grandfathering) or through a compensation system (sale, auction, etcetera).

When the certificates are allocated and the system starts to work, pollutants are free to negotiate their certificates as in the "pure market". Participants act according to their strategies and interests while reducing and meeting the objectives of GHG reductions. Whenever the unit price of a certificate is greater than the cost of reducing pollution by the same unit, the company will reduce polluting emissions, and it would seek to sell the rights to a company with higher costs, in order to make some profit. This will gradually reduce the price of the certificate issuance. By contrast, a polluting company, whose abatement costs are higher than the price of allowances, will seek to buy certificates in the market, pushing the price to a higher value (Wagner 2013).

Finally, at the end of each period, companies are asked to report their total emissions, and the performance of this instrument is simply assessed by comparing the emissions each company has against the certificates that they hold. If the company exceeds their allowances, an additional compensation (fine) per tonne is claimed.

Wagner (2013) summarises that an environmental policy on climate change based on a cap-and-trade system should have the next three objectives:

1. To limit GHG emissions and to achieve the same reduction to prevent dangerous effects of global warming as the best available scientific sources indicate.
2. To minimise the costs of reducing GHG as a whole through the markets.
3. To encourage investment in eco-efficient, clean and low-emission processes and technologies: ultimately, the price signal of certificates would be one of the references to stimulate such investments.

4.3 Principles for Designing a Cap-and-Trade System

In an analysis regarding the experience with various cap-and-trade systems in the world, Betsill and Hoffmann (2011) contextualise the design of the instrument and point out that those systems that have proven to perform well have certain characteristics in common. While the design principles vary depending on the regulatory framework, they generally tend to be flexible and have different options to achieve the objectives outlined before.

For this analysis, the EU ETS was primarily used to establish the main design principles that a possible cap-and-trade system in Mexico should consider.

The scope of regulation was the first principle identified. In the context of the policy, the scope must have two parameters: (i) the magnitude of GHG reductions and (ii) the number of emission sources, which would be subject of regulation. Both parameters are dependent: the decision on the size of the reductions has an impact on the number of sources that should be regulated and vice versa. The choice of these parameters is also linked to the economy of the region, for instance, the greater the emission reductions and the shorter the time to achieve them, the greater the economic cost. Therefore, the choice of both parameters is a complex act of political and economic balance (Burtraw et al. 2010). Furthermore, Bressers and Huitema (1999) stress the role of institutions and interest groups in the policy-making process rather than just take into account the cost-effectiveness assumption of market-based instruments.

The second principle focuses on the characteristics of the cap in every phase of operation. Wagner (2013) and other authors state that a cap on emissions that decreases over the time is a key element in the design, since knowing the path of the cap in advance allows companies to plan their investment decisions. That said, the cap should not only be based on climate science but also on the fairness of the amount each sector needs to reduce. Once again weighing the level of what can be considered fair is far from easy since priority-setting in policy-making is a multi-actor interaction process (Bressers and Huitema 1999).

The third principle is based on how permits are initially allocated. Tietenberg (2003) proposed that the allocation can be carried out through an auction or based on the historical emissions over the past years for free (this process is known in the economic jargon as grandfathering). From there, each sector can buy and sell permits depending on their own ability to reduce emissions. If emission certificates are issued free of charge to the pollutants, then they obtained the right to pollute to a negative cost; however, if they are auctioned by the government, polluters pay for the right to emit one tonne (Betsill and Hoffmann 2011). This principle has been proven to be the most controversial one since cases of overallocation, like in the EU ETS, have affected the price signal of the certificates, pushing it to lower levels.

The fourth principle identified was the point of regulation, that is, who is subject to participate in the system. The main options include an upstream approach, which would require regulation in the fuel and energy production and the downstream approach, which imply that users of fossil fuels and energy would participate. The first approach would naturally include fewer entities to regulate and the second a larger amount of covered entities (Hargrave 2000). There are sectors in which selecting the point of regulation is such a challenge in terms of implementation. For instance, in the transport sector, a downstream approach involves an enormous amount of emission sources, and on the other side in an upstream approach, the number of sources is limited or simply belongs to another jurisdiction or country (Burtraw et al. 2010).

Flexibility is the fifth principle of design, and this can be presented in different measures. For instance, the scheme "save and borrow" allows the certificate prices to be more stable.[1] Another scheme that can help to moderate the price volatility of certificates at strategic moments is the "allowance reserve", which consists of a portfolio of allowances that is not distributed to the companies immediately. The "offsets" are another possible mechanism for flexibility, in which it is allowed to compensate the value of the certificates through another instrument such as the Clean Development Mechanisms (CDM) or the joint implementation (JI) credits. Flexibility is key in controlling costs (Wagner 2013).

Finally, the sixth design principle recommends an adequate supervision. It is desirable that the authorities try to avoid monopolistic practices in the market for emissions or showing other forms of anticompetitive or abusive behaviour. This is clearly not unique to the cap-and-trade system, which is similar to the regulatory oversight of any market. It's also important to ensure that all the participants subject to the cap-and-trade system are responsible for fulfilling their obligations to reduce emissions and face severe penalties in case they don't meet the regulation.

Table 4.1 compiles the six design principles presented throughout this section and the main practices and approaches that have been considered in various systems in the world.

4.4 Feasibility Analysis of a Cap-and-Trade System in Mexico: Contextual Analysis

In this section, the authors look at the relevant factors playing a role of the cap-and-trade system, such as the (Sect. 4.1) potential sectors eligible for the cap-and-trade implementation and (Sect. 4.2) the Mexican context comparing with other market-based instruments.

---

[1] The "savings" refer to the possibility that companies can save unused permits for future periods, while the "borrowings" represent the possibility of asking permission to use future allocations in the current period.

Table 4.1 Design principles of a *cap-and-trade* system (Own contribution)

Design principle	Approaches
Scope of regulation	Magnitude of GHG reductions
	Number of emission sources which are subject to regulation
	Situation of the sectors in the region (barriers)
Cap characteristics	It should decrease over time
	It should be predictable in its path
	It should be calculated in terms of climate science
	It should be consistent
Allocation	Auction
	Grandfathering
	Hybrid
Point of regulation	Upstream
	Downstream
	Hybrid
Flexibility	"Save" and "borrow" schemes
	Additional offsets like CDM and JI
	Binding schemes with other cap-and-trade systems
	No leakage schemes
Supervision	Anticompetitive practices
	Dishonest practices

4.4.1 Potential Role of a Cap-and-Trade System in the Climate Change Policy in Mexico and Potential Sectors for Implementation

Although Mexico is not among the members of Annex I countries of the Kyoto Protocol (KP), the climate change policy of the country has made notable progress in recent years to outline a legal framework to address the problem of global warming. The most outstanding result came in June 2012 when the General Law on Climate Change was published.

Particularly, the use of an emissions trading system such as the cap-and-trade was introduced for the first time as a possible action plan to lead part of the climate change policies. One motivation for this paper arises from the articles 94° and 95° from Title V, Chapter IX of that law which state:

> Article 94. "The Secretariat, with the participation of the Commission and the Council may establish a voluntary emissions trading system in order to promote emission reductions that can be carried out with the least possible cost, measurable, reportable and verifiable". (Ley General de Cambio Climático 2012, p. 37)
>
> Article 95. "Those interested in participating voluntarily in emissions trading may carry out operations and transactions linked to trading systems from other countries, or may be used in international carbon markets under the terms provided by the legal provisions that are applicable". (Ley General de Cambio Climático 2012, p. 37)

Although this law establishes the legal feasibility of developing a national system of emission trading as an economic mitigation instrument, it is important to underline that its operation does not arise mandatory, in line with Mexico's place in international agreements on climate issues. This law provides as well very explicit (but not binding) emissions reduction targets in the medium and long term: a 30% reduction compared to a business-as-usual scenario in 2020 (960 $MtCO_{2\text{-}e}$) and a 50% from 2000 levels by 2050 (Ley General de Cambio Climático 2012).

A final national framework that raises the feasibility of developing a market for emission certificates is given in the National Climate Change Strategy, which describes the strategic priorities and lines of action to follow from a federal level to a municipal level. Through pillar number two of the strategy (points P2.1 and P2.15), it promotes including economic, fiscal and financial markets in climate change policies to encourage mitigation and adaptation, citing, for example, the voluntary carbon markets (Secretaria de Medio Ambiente y Recursos Naturales 2013).

While the above references provide support and legal feasibility of implementing a system such as the cap-and-trade, there are other factors that could oppose such implementation.

Regarding the potential sectors for implementation, according to the report of the International Energy Agency (2013), Mexico contributed in 2011 to 1.4% of global emissions of CO_2, ranking 12th with the highest emissions. This contribution does not seem to be significant when compared with the large emitters; however, considering that Mexico is a country with high growth perspectives for the coming years coupled with the fact that the demand for energy from conventional sources will still grow, then mitigation of GHG should be a commitment that Mexico could tackle according to their possibilities and realities.

According to the National Institute of Ecology and Climate Change (INECC) in 2010, there were emitted 748 $MtCO_{2\text{-}e}$ to the atmosphere, which represented an increase of 33% compared to the emissions of 1990. As shown in Fig. 4.1, CO_2 emissions in 2010 are still the largest source of GHG. Therefore, designing a

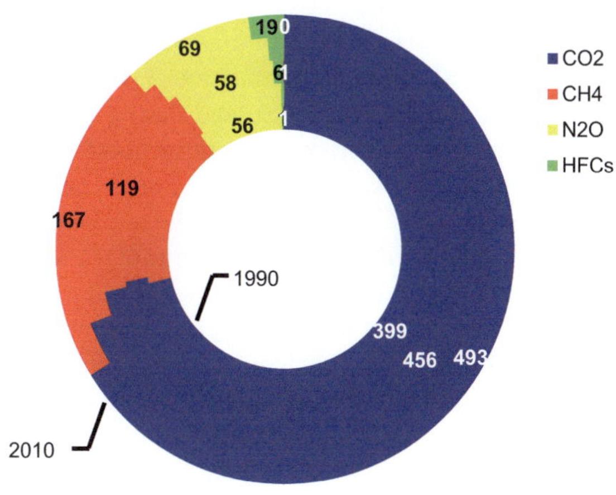

Fig. 4.1 Distribution of GHG emissions in Mexico in 1990, 1995, 2000, 2005 and 2010 (million $CO_{2\text{-}e}$ tonnes; Adapted from INECC 2013)

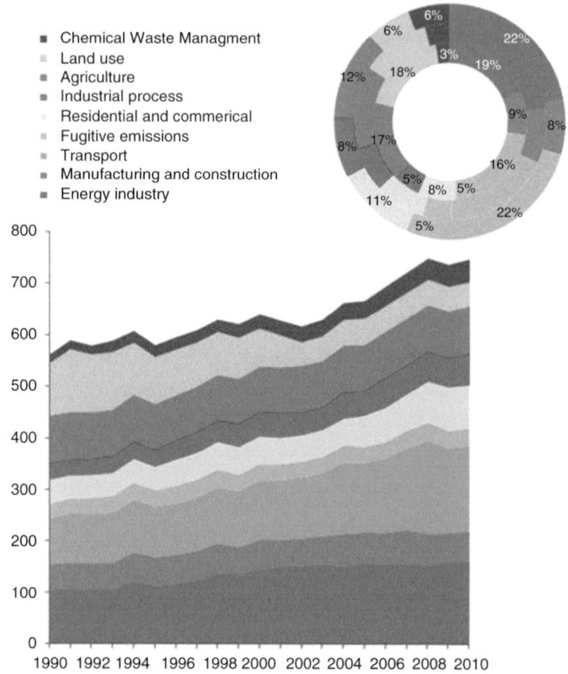

Fig. 4.2 Evolution of GHG emissions by sector (million $CO_{2\text{-e}}$ tonnes; Adapted from INECC 2013) and distribution by sector in 2000, 2005 and 2010 (the inner circle is the oldest year)

cap-and-trade system whose scope of regulation should focus on CO_2 covers a wide range. In comparison, the EU ETS operates only for CO_2 emissions because of the percentage represented by this gas and also because more accurate estimations can be done for monitoring (Fig. 4.2).

With this first evidence, the energy industry in general and the industrial processes would be likely areas to implement a system of emissions trading.

In the currently operating cap-and-trade systems in other countries, the sectors mentioned above are covered by the scheme due to similar reasons. For the EU ETS, the following industries are considered subject of regulation: combustion plants, oil refineries, steel plants, cement, glass, brick, lime, ceramics, pulp, paper and since 2013 airlines (European Commission 2013). EU ETS initially operated only for electricity generation facilities, and during the second phase of implementation, other sectors were incorporated to cover more than 11,500 installations throughout the continent. In Mexico, a similar case could be applied, since the electricity industry represents almost a quarter of total emissions.

It is noteworthy that back in 2012 one of the amendments of the General Law on Climate Change urged the government (federal, state and municipal) to develop GHG inventories, and since August 2015, companies that exceed 25,000 $tCO_{2\text{-e}}$ are required to submit their annual emissions report through the National Registry of Emissions (RENE). Due to the recent application of this tool, which undoubtfully will help to create a more certain national inventory, for the purposes of this paper, the analysis was only limited to the information on the Registry of Emissions and

Table 4.2 CO_2 emissions and number of facilities from nonmobile sources in 2012

	Emissions ($MtCO_2$)	Facilities reported
Energy	**226,6**	**1547**
Power generation	122,2	115
Power facilities from CFE	*52,3*	*68*
Others	*69,9*	*47*
Oil and petrochemical	37,2	266
Pemex	*34,7*	*219*
Others	*2,5*	*47*
Metal and steel industry	29,9	347
Chemical	29,5	514
Car industry	3,1	203
Pulp and paper	2,5	78
Glass	2,1	24
Industrial process	**28,5**	**124**
Cement and lime	25,5	69
Chemical waste management	1,1	35
Others	1,7	20
Others not classified	**2,9**	**223**

Adapted from INECC (2013)

Transfer (RETC). In Table 4.2, CO_2 emissions from different sectors are summarised by the number of facilities.

In 2012, 1894 facilities emitted around 258 $MtCO_2$ into the atmosphere, and only 5% of them (100 facilities) accounted for almost 81% of the total reported. Among those 100 facilities, 53 belong to the electricity generation sector and 15 to the oil sector. In Fig. 4.3, it can be seen that the contribution of CO_2 emissions depends on the number of registered facilities. This figure likewise emphasises that a relatively small number of facilities have the highest concentration of emissions.

The measure mentioned in the preceding paragraph allows establishing a threshold in which the largest number of emission sources is considered without compromising those whose activities are very low. For the case of the EU ETS, all facilities are included; in other words there is no threshold, but in other systems, for example, in California, USA, a threshold of 100,000 metric tonnes is established, or in the case of the regions of Canada, the threshold is 25,000 metric tonnes.

As it can be seen in Fig. 4.3, a large number of facilities in Mexico do not even represent 5% of total emissions. This could represent both an opportunity and a barrier to implement a cap-and-trade system, opportunity in the sense of fewer regulated facilities and the lower costs for verification and monitoring and a barrier since having few participants could reduce market liquidity for the emission certificates.

Parallel, different reports and articles like Morris (2009), Burtraw et al. (2010), IETA (2013) and the European Commission (2010) indicated that Mexico has a great potential for reducing emissions in the sectors of electricity generation, fuel refining, cement and metals.

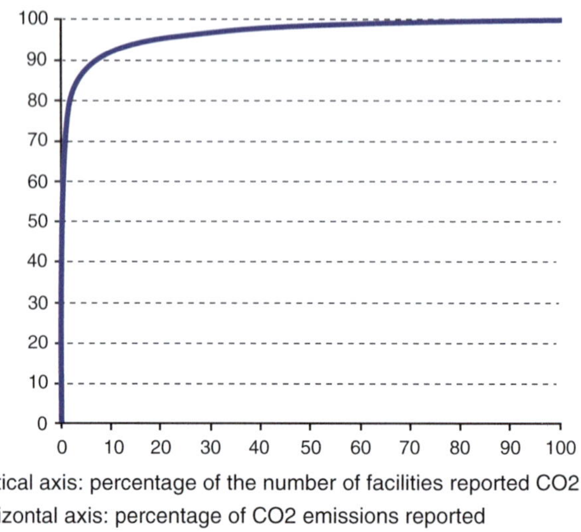

Fig. 4.3 Scatter of the number of facilities regarding their CO_2 emissions. (Adapted from INECC 2013)

4.4.2 Mexico's Context with Other Market-Based Instruments

Mexico has positioned itself as one of the favourite countries to host several CDM projects. According to official figures at the end of 2013, it had accumulated 207 projects, ranking fifth issuing just over 22 million reduction certificates (CERs).

In several reports, including a study by Centro Mario Molina (2008), it is concluded that Mexico still has not yet exploited the full potential of investment in CDM projects. These studies indicated that the country could had reached up to 100 million tonnes of reduced $CO_{2\text{-e}}$ per year only in the energy sector, of which power generation and the oil sector could represent at least 50%.

Kleper (2011) concluded that under the current scheme, the willingness of developing countries to agree with a cap on their emissions would be limited since their efforts to reduce would be accounted for industrialised countries and not for them due to the fact that most of the financial resources come from them. Using CDM as an offset in post-Kioto agreements would require industrialised countries to make more robust reduction commitments.

In parallel, a continuous slump in the demand of these mechanisms has been observed since 2013, especially in the third phase of the EU ETS where CERs from Mexico were no longer allowed (European Commission 2013). That is why Kepler (2011) and some other authors suggest that a scenario is more likely wherein CDMs completely disappear or are totally transformed.

That said, introducing a domestic cap-and-trade in Mexico would not cause any conflict with the current situation of CDM in the world and more precisely in the country.

On the other side, on December 2013, the incorporation of an "environmental tax" in Mexico's tax system was approved which can be addressed by cash or by carbon credits obtained for projects in energy efficiency developed in Mexico and endorsed by United Nations (UN) (Presidencia de la República 2013). The carbon tax was set at a "low" level of 5.70 USD/tonne of $CO_{2\text{-}e}$ in order not to create negative effects on the economy in the short term. However, it will increase gradually each year to generate more resources for mitigation and adaptation to climate change (according to UN figures, the costs for mitigating climate change vary from 20 to 25 USD/tonne of $CO_{2\text{-}e}$).

Moreover, Mexico declared its preference over carbon tax by stating that it was the most straightforward market-based instrument to apply in the country since there was a great concern about the impact of operating costs, supervision, monitoring and the entire institutional infrastructure that for an emission trading system would need to be implemented.[2]

In this context, using a carbon tax as a market-based instrument does not exclude the possibility of a joint implementation with another mechanism such as the cap-and-trade. For instance, in Europe, a number of countries including Sweden, Norway, Denmark and the UK have introduced carbon taxes besides the current EU ETS in operation. In addition, many members of this system (especially Germany and Spain) provide other policies such as subsidies for electricity generation from renewable sources. Or even more, in the UK, there is an additional cap-and-trade system for the service and public sectors (Fankhauser et al. 2011).

Fankhauser et al. (2011) also argue that a hybrid policy, such as putting a tax as a ceiling price on a cap-and-trade scheme, has major economic and environmental benefits rather than applying them separately. In such system, it would require companies to either pay prespecified taxes or simply meet the required permissions but not both.

Therefore, it is concluded that in the Mexican case, within a context of recent incorporation of carbon taxes and because the level is now considered "low", it is also possible that both instruments can interact together since it could send a stronger price signal and could give more certainty to the level of emissions reduction.

Additionally, the Energy Transition Law, which was published in December 2015, will establish different mechanisms to achieve the use of "clean energies" in at least 35% of Mexico's energy mix. Among these mechanisms, Clean Energy Certificates will be generated by 2018–2019, which could potentially be linked to a cap-and-trade national system.

[2] For more information regarding the differences between carbon taxes and emission trading systems, refer to Bristow et al. (2010), Xiangsheng et al. (2013), Balderas (2012), Chesney et al. (2013) and Aghion et al. (2009).

4.5 Feasibility Analysis of a Cap-and-Trade System in Mexico: MACC Assessment

Governments in their quest to reduce CO_2 emissions profitably and cost-effectively have different tools available for making decisions, and one of them is the use of the marginal abatement cost curves (MACC).

The MACC are used by different actors for different purposes. Regulators, for instance, use them to identify where interventions can be effective in certain public policies (e.g. establish a green tax or cap-and-trade system) and at what cost. Companies can use them to guide their long-term investment decisions to reduce emissions and improve competitiveness, while traders generally use them to derivate the function that provides the price of emission certificates.

Kesicki and Strachan (2011) point out that it is noteworthy that the MACC have been used worldwide as a standard tool to illustrate the economics of climate change mitigation. As an example of this, we have the case of the UK, who has based some of his policy decisions using MACC.

A MACC can, in a simple way, give a price approximation of the certificates associated with a reduction percentage and the potential of applying certain policies such as the cap-and-trade. This is based on the logic that any polluting mitigation option that is below such price would be implemented because they are economically appropriate (Kesicki 2011).

In Mexico, the Mario Molina Centre in cooperation with McKinsey Co. (2008) built the first MACC with horizon to 2030, which served to establish some of the specific objectives of the Strategic Plan for Climate change (CICC 2009). Five years later, the US Agency for International Development (USAID) presented an updated version of the MACC and concluded that Mexico has a potential for reducing its GHG emissions by 33% with respect to the levels of a business-as-usual scenario for 2020 (USAID 2013). It also shows that 87% of the reduction potential is concentrated in five sectors, the most important transportation, power generation, oil and gas and forestry. This particular curve is taken as reference to assess the need for a cap-and-trade system in this paper. Figures 4.4 and 4.5 show how to read the curve and which sections are more suitable for certain policies.

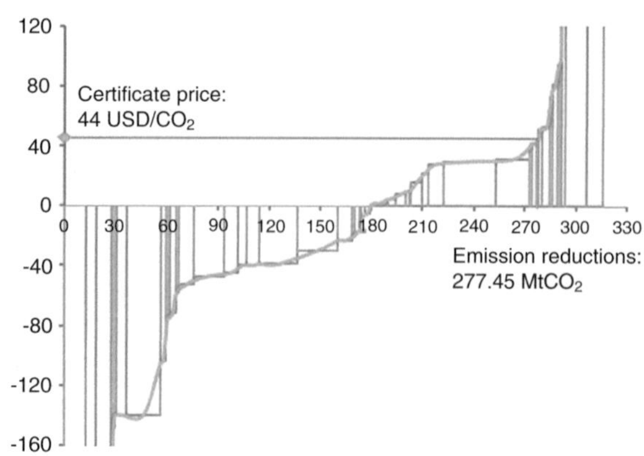

Fig. 4.4 Setting the price of the emission certificates in a *cap-and-trade* system

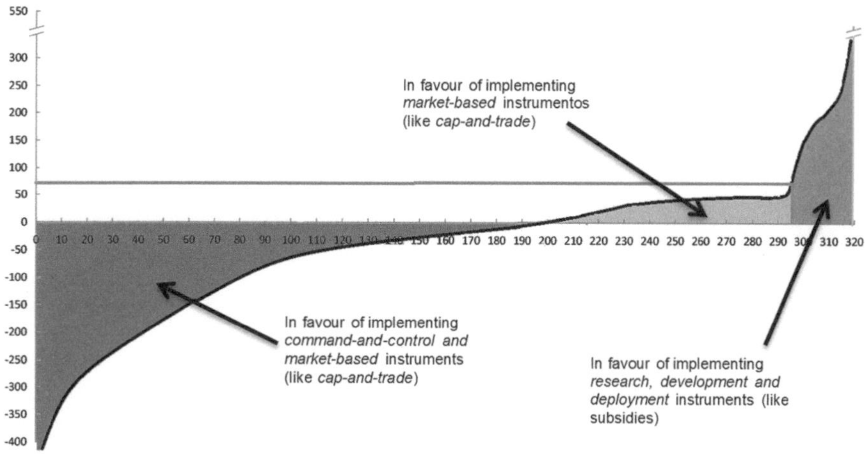

Fig. 4.5 Need for a cap-and-trade system through the reading of a MACC

In the following paragraphs, the results obtained from the assessment of Mexico's MACC are going to be presented. Data for calculating the potential of introducing a cap-and-trade system as well as the costs associated with it were obtained from the latest update of USAID's report entitled "Updated analysis on Mexico's GHG baseline, marginal abatement cost curve and project portfolio". A total of 67 mitigation options from an universe of 126 were obtained. Those that were not reported were not included in this study (lack of data). However, these 67 mitigation options accounted for 91% of the total potential.

The sectors included were agriculture, forestry, cement, chemical, metallurgical and steel, oil and gas, power generation, transportation and waste management. The costs for implementing the various mitigation options were calculated using McKinsey Co. methodology.[3] Since there are some interactions between the mitigation options, the order in which these options are carried out may affect the total reduction potential.

According to the processed information, the reduction potentials are concentrated in four sectors: forestry, transportation, power generation and oil and gas. In Table 4.3, the most important results are condensed.

The percentage of measures with negative cost almost accumulates 57% of all measures; in other words, more than a half of the mitigation options not only achieve a significant reduction in GHG emissions but also save money in the long run. The average cost of all options results in a financial gain of 30.2 USD/tCO_2, which could translate into a benefit for the economy as a whole by almost 1% of GDP by 2020.

[3]The social discount rate used in this method was 4%, and the following costs were not included: transaction costs, information or communication costs, subsidies for taxes and costs of monitoring implementation. For further information on McKinsey Co. methodology for cost estimations, refer to USAID (June 2013).

Table 4.3 Results from all mitigation options

Sector	Abatement potential ($MtCO_{2\text{-}e}$)	Abatement potential (% BAU scenario)	Average cost mitigation options (USD/tCO_2)	Total cost-benefit (million USD)	Total benefit cost (as GDP %)	Negative cost mitigation options (% of total potential)
Agriculture	18.9	−13	−17.7	−335.3	−0.033	50
Forestry	73.0	−145	31.0	2266.3	0.221	0
Cement	1.4	−8	−139.1	−194.7	−0.019	100
Chemical	5.3		−4.5	−23.6	−0.002	14
Metallurgical and steel	4.9		−36.9	−180.7	−0.018	88
Other industries	7.4		−40.0	−296.0	−0.029	100
Oil and gas	47.0	−45	−116.6	−5481.8	−0.534	91
Power generation	55.0	−36	−44.0	−2421.5	−0.236	64
Transportation	55.0	−20	−84.2	−4632.5	−0.451	85
Waste management	44.0	−61	11.1	489.0	0.048	70
Total	**311.9**	**−32**	**−30.2**	**−9530.8**	**−0.93**	**57**

Adapted from USAID (2013)

This also indicates that there are sectors, which find more profitable (but not easy) to reduce their emissions because it also represents savings in costs.

The forestry sector has the biggest potential, which represents 23% of the total. Implementing all mitigation options could even turn it into a net sink. However, it also represents the sector with the greatest impact in terms of costs, because 100% of the options have positive cost, implying that they are mostly absorbed by society. The estimated average cost of this sector is at 31 USD/tCO_2, almost 9 times higher than the current carbon tax and marginally below the maximum value observed in the carbon market in the EU.

Transportation is the second largest potential sector (18%). Eighty-five percent of all the options have a negative cost; therefore, the average cost results in a financial gain of around 84 USD/tCO_2 or in terms of GDP 45 basis points.

The electricity generation sector shares the same position as transportation, accounting also 18% of the total. Sixty-four percent of the options have negative cost, bringing the average cost to a financial benefit of 44 USD/tCO_2. In terms of GDP, the benefit translates into 0.2%.

For the oil and gas sector, a potential reduction of 55 $MtCO_2$ was obtained, representing 15% of the total. This particular sector has the highest proportion of negative cost options, 91% of the total, as well as in terms of financial gain, which account over 0.53% of GDP by 2020. The average profit per tonne of CO_2 is around USD 116.

Finally, to determine the need for a cap-and-trade system in Mexico, two references or scenarios were used:

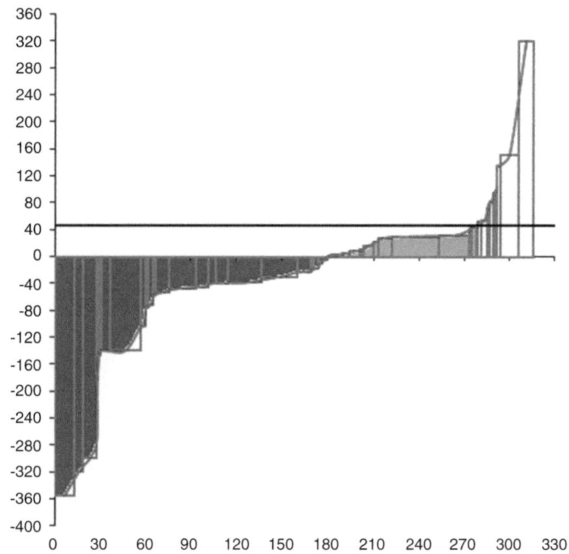

Fig. 4.6 Potential use of a *cap-and-trade* system in Mexico throughout the economy. (Adapted from INECC 2013)

1. The maximum price of EUAs (EU ETS certificates): 45.69 USD/MtCO_2 (the exchange rate considered was 1.3846 EUR/USD).
2. The tax value of the fuel most used in Mexico's energy mix: 3.53 USD/MtCO_2 (the exchange rate considered was 13.00 MXN/USD).

As a first step, a cap-and-trade for the economy as a whole was addressed. In the first scenario, if the equilibrium price was at 45.69 USD/tCO_2, the potential supply of certificates could reach 280 million by 2020, and the percentage of potential options which would be appropriate with a cap-and-trade instrument accounts for 89%: 57% with negative costs and 32% with positive costs. That said, a cap-and-trade system throughout the economy has a high potential for implementation; however, the percentage of options that represent negative net costs (or a win-win in the economy) is almost as twice as the percentage of positive costs. In that sense, introducing a cap-and-trade system could generate more supply than demand, which is good in a sense of reductions, but counterproductive in the price of emission certificates in the long term. Figure 4.6 shows, firstly, the level of reduction that could be achieved, given an equilibrium price of 45.69 USD/tCO_2 (denoted by the black line), and, secondly, the regions of the options which are appropriate for a cap-and-trade system (options with positive cost are marked in light gray and options with negative cost are marked dark gray).

Although the potential of those options is appropriate to a cap-and-trade scheme, the next step is to determine if there is predominance in a few sectors. To do this, all mitigation options that belong to the 89% previously calculated were analysed. Table 4.4 shows the potential of implementing a cap-and-trade system for each sector. They are prioritised according to the potential supply of certificates.

Table 4.4 Summary of the potential of introducing a *cap-and-trade* system in various sectors under an equilibrium price of 45.69 USD/tCO_2

Sector	Potential supply of certificates (million)	Potential options of being considered in a *cap-and-trade* (%)	Potential options with positive cost in a *cap-and-trade* (%)
Power generation	52.5	95	88
Gas and oil	44.5	95	33
Transportation	47	85	0
Waste management	32	73	8
Agriculture	18.1	96	91
Forestry	15.5	21	21
Other industries	7.4	100	100
Metallurgical and steel	4.3	88	0
Chemical	4.2	81	86
Cement	1.4	100	0
Total	**280**	**89**	**77**

Adapted from INECC (2013)

Table 4.4 shows that the power generation sector, the oil and gas industry as well as transportation are the sectors in which the implementation of a cap-and-trade with initial price of 45.69 USD/tCO_2 not only has the greater chance of implementation but also the greatest reduction potential.

To complete the analysis, the potential for implementation was also assessed considering the current level of Mexico's carbon tax (3.53 USD/tCO_2). The potential supply of certificates could reach 189 million units by 2020, which is 48% lower compared to the maximum price of certificates in the EU ETS. The percentage of potential options that would be appropriate with a cap-and-trade system decreases to 60%: 57% with negative cost and only 3% with positive cost. That said, a cap-and-trade system across the economy would have a considerable potential for implementation; however, the percentage of options that represent negative cost account for almost the entire potential.

As for the comparison by sector, implementation rates are basically the same for oil and gas industries and transportation. For the electricity generation, the potential decreased almost 20% but still remains at a high rate of implementation. All this would suggest that these three sectors, even considering an initial lower equilibrium price, might be technically feasible to implement a cap-and-trade system.

4.6 Implications of Circular Economy into Cap-and-Trade Systems

In contrast to cap-and-trade instruments, circular economy thinking establishes a more holistic approach since methodologies like life cycle assessment lead not only to emission reductions but also to awareness of the raw materials used and the effects of products' disposal. Even though this thinking has influenced a considerable amount of policies in the EU, there has been no explicit integration into climate mitigation policies.

Niederberger et al. (2013) explore the convergences and divergences that market-based policies and circular economy thinking have and through a series of case studies present interesting implications of how market-based mechanisms can be designed to drive, instead of hindering, the transition to a low-carbon economy. The authors, firstly, indicate that the paradigm of market-based instruments focuses on the flexibility in fulfilling GHG mitigation through the least cost options, while the paradigm of circular economy (CE) shifts from waste management to resource management, leading to the most productive use of resources. Secondly, they expect that the outcomes for market-based instruments are directly linked to emission reductions that address environmental issues like global warming, ozone depletion and human toxicity. In the case of CE, because it applies the life cycle approach, the expected outcomes are the substitution of material inputs by recovered, reused, recycled resources at any stage of the supply chain.

Finally, Niederberger et al. mention that the boundary considerations of market-based policies may vary from country, region and scope of regulation, while CE thinking does not have geographical constrains and considers all life stages of a material/product in any sector.

Since life cycle assessments generally evaluate "potential" impacts and market-based instruments try to represent "real and verifiable" emissions reductions, there is a notable inconsistency between these two thinkings. Even further, from Niederberger et al. (2013), it was shown that there are some deficiencies in market-based instruments when they attempt to stimulate the transition to efficiency in the use of resources. It was argued that the carbon price signal, which is the main direction for participants in a cap-and-trade system, has only marginally changed the use of efficient technologies. In Germany, for instance, there is no evidence that the cost increment in electricity due to EU ETS has translated into a strong signal for households to purchase efficient appliances.

Niederberger et al. (2013) also addressed that market-based instruments until now have failed to stimulate the adoption of cradle-to-cradle practices. For example, it was found that most manufacturers have not managed to generate revenues from offset activities related to cap-and-trade flexibility mechanisms. That said, manufacturers do not really see the economic benefit of product innovation through this system.

Finally, recycling and reusing are key principles for circular economy, and therefore a sound market-based policy should take into consideration these basic processes. However, carbon markets do not directly reward pure recyclers, because the service they provide results in avoidance of upstream GHG emissions reductions associated with raw materials needed to manufacture final products downstream, activities that are beyond the control of the recycling company. There is also no carbon market mechanism that makes manufacturers responsible for carbon embedded in the raw materials they use (Niederberger et al. 2013).

4.7 Conclusions

Mexico has positioned itself as a leader in the emergent economies to address climate change. Through the General Law of Climate Change and the National Strategy of Climate Change, it has endorsed this commitment.

This paper sought to answer if it is feasible to implement a cap-and-trade system in Mexico as part of its climate policy. This scheme aims to limit and reduce GHG emissions in the country and has certain advantages compared with other policies. Even in certain regions, such as the European Union, the cap-and-trade system is considered the backbone of the climate policy. However, recent events and certain elements in its design have shown how fragile any market instrument can actually be, mainly in a context of financial crisis and increasingly polarised negotiations on the international agenda.

As part of the analytical framework to test the research hypothesis, the marginal abatement cost curves (MACC), whose results proved partially that a cap-and-trade system is a feasible option in Mexico, were used. This is because the potential for implementation and associated costs to reduce emissions makes it suitable for the development of a market for emission certificates. However, there are certain barriers and contextual elements that could undermine the effectiveness and thus minimise the fundamental objectives of the scheme.

Firstly, the evidence displayed regarding Mexico's context with other market-based instruments points out that a scenario where CDMs are redesigned is more likely to occur since other ETS (like the third phase of the EU ETS) are no longer allowing CERs from specific countries like Mexico. Therefore, it was concluded that the CDM does not represent a potential barrier to the introduction of a cap-and-trade. In the case of the recent implementation of a carbon tax in Mexico, it is possible that the cap-and-trade system and this tax can interact together. This is mainly because the current level of tax is eight times less than the maximum that has been reported in European markets, as well as the evidence of joint policies in other countries. Therefore, it was concluded that the carbon tax is not a potential barrier for the cap-and-trade system. Furthermore, it could create a hybrid instrument that sends a stronger price signal, giving more certainty on the level of emissions that could be reduced.

It was also found that the percentage of measures to reduce GHG with negative cost agglomerates 89%, of which 57% of the options have negative cost. It could not only mean a significant reduction in Mexico's emissions but also could translate into a benefit for the economy as a whole by almost 1% of GDP by 2020. It was also identified that the sectors of power generation and the oil and gas industry are the sectors in which the implementation of a cap-and-trade would have the greatest chance of implementation.

Finally, it is worth mentioning that since Mexico is characterised to be a manufacturing economy, a possible cap-and-trade system in the country should be designed rewarding more the most productive use of resources, rather than any marginal emission reduction. This would align carbon markets signals to promote better material management throughout its life cycle.

References

Aghion P, Hemous D, Veugelers R (2009) No green growth without innovation. Bruegel, Bruselas

Balderas Torres A (2012) Yes in my backyard: market based mechanisms for forest conservation and climate change mitigation in La Primavera, México, PhD thesis. University of Twente, Netherlands. http://doc.utwente.nl/82280/. Accessed 5 Feb 2014

Betsill M, Hoffmann M (2011) The contours of "cap-and-trade": the evolution of emission trading systems for greenhouse gases. Pol Res 28:83–106

Bressers H, Huitema D (1999) Economic instruments for environmental, protection: can we trust the "magic carpet"? Int Polit Sci Rev 20:175–196

Bristow A, Wardman M, Zanni A, Chintakayala P (2010) Public acceptability of personal carbon trading and carbon tax. Ecol Econ 69:1824–1837

Burtraw D, Kopp R, Morgenstern R (2010) Feasibility assessment of a carbon cap-and-trade system for Mexico. RFF report, Resources for the future

Centro Mario Molina, McKinsey Co (2008) Low carbon growth. A potential path for Mexico. Centro Mario Molina para Estudios Estratégicos de Energía y Medio Ambiente, Mexico City

Chesney M, Gheyssens J, Taschini L (2013) Environmental finance and investments. Springer-Verlag, Zurich

CICC (2009) Programa Especial de Cambio Climático. (DOF - 28 de agosto de 2009)

European Comission (2010) Global sectoral study: final report. http://ec.europa.eu/enterprise/policies/sustainable-business/climate-change/sectoral-approaches/index_en.htm. Accessed 18 Feb 2014

European Commission (2013) The EU Emission trading system: factsheet. Bruselas: European Commission. Publication Office

Frankhauser S, Hepburn C, Park, J (2011) Combining multiple climate policy instruments: how not to do it. Centre for Climate Change Economics and Policy No. 48

Hargrave T (2000) An upstream/downstream hybrid approach to greenhouse gas emissions trading. Centre for Clean Air Policy, Washington DC

IEA (2013) CO_2 emissions from fuel combustion. International Energy Agency Press, Paris

IETA (2013) The world's carbon markets: a case study guide to emissions trading in Mexico. International Emission Trading Association. http://www.ieta.org/worldscarbonmarkets. Accessed 27 Nov 2013

INECC (2013) Inventario Nacional de Emisiones 2010. SEMARNAT. http://www.inecc.gob.mx/cpcc-lineas/1165-inem-1990-2010. Accessed 12 Feb 2014

Kesicki F (2011) Marginal abatement cost curves for policy making – expert-based vs. model-derived curves. UCL Energy Institute, University College London, London, pp 1–19

Kesicki F, Strachan N (2011) Marginal abatement cost (MAC) curves: confronting theory and practise. Environ Sci Pol 14:1195–1204

Kleper G (2011) The future of the European emission trading system and the clean development mechanism in a post-Kyoto world. Energy Econ 33:687–698

Lascoumes P, Le Gales P (2007) Introduction: understanding public policy through its instruments – from the nature of instruments to the sociology of public policy instrumentation. Governance 20:1–21

Ley General de Cambio Climático (2012) Cámara de Diputados del H. Congreso de la Unión. http://www.diputados.gob.mx/LeyesBiblio/pdf/LGCC.pdf. Accessed 20 Apr 2013

Morris JF (2009) Combining a renewable portfolio standard with a cap-and-trade policy: a general equilibrium analysis (Master thesis). Massachusetts Institute of Technology, United States. http://citeseerx.ist.psu.edu/viewdoc/download?rep=rep1&type=pdf&doi=10.1.1.208.6312. Accessed 10 Mar 2014

Niederberger AA, Shiroff S, Raahauge L (2013) Implications of carbon markets for implementing circular economy models. Eur J Bus Manag 5:187–199

Presidencia de la República (2013) Iniciativa de Decreto por el que se reforman, adicionan y derogan diversas disposiciones de la Ley del Impuesto al Valor Agregado, de la Ley del Impuesto Especial sobre Producción y Servicios y del Código Fiscal de la Federación. Gobierno Federal, Mexico City

Secretaría de Medio Ambiente y Recursos Naturales (2013) Estrategia Nacional de Cambio Climático. Visión 10-20-40. Gobierno Federal, Mexico

Simmons B, Elkins Z (2004) The globalization of liberalization: policy diffusion in the international political economy. Am Polit Sci Rev 98(01):171–189

Tietenberg T (2003) The tradable permits approach to protecting the commons: lessons for climate change. Oxford University Press, Oxford

USAID (2013) Updated analysis on Mexico's GHG baseline, marginal abatement cost curve and project portfolio. http://www.mledprogram.org/docs/MLED-DOCU-Updated-analysis-Mexicos-GHG.pdf. Accessed 23 Feb 2014

Wagner G (2013) Carbon cap and trade. In: Shogren JF (ed) Encyclopedia of energy, natural resource, and environmental economics. Elsevier, Amsterdam, pp 1–5

Xiangsheng D, Jingjing X, Zenglong Y (2013) Policy design and implementation issues of regulating. Intl J Environl Sci Dev 4(3):321–326

Chapter 5
Circularity of Wastes: Stakeholders Identity and Salience for Household Solid Waste Management in Cimahi City, West Java Province, Indonesia

Vina Septi Suherman, María-Laura Franco-García, Oekan S. Abdoellah, Denny Kurniadie, and Yuli Astuti Hidayati

Abstract This paper presents the operationalisation of stakeholders identity and salience theory (Mitchell RK, Agle BR, Wood, DT, Acad Manag Rev 22(4):853–886, 1997) with the purpose to classify the stakeholders involved in the household solid waste management in Cimahi City, West Java Province, Indonesia. This classification will benefit circularity of solid waste management strategies that involve diverse actors needing to collaborate. Several studies have been conducted to identify the stakeholders in solid waste management, but none has been carried out yet to classify their salience in the systematic approach described by Mitchell's theory. Such classification aims to draw a line between the stakeholders who play a vital role in the household solid waste management process and those who have minor contributions towards the process. Hence, the research questions aligned to this aim are: (1) Who are the stakeholders of household solid waste management in Cimahi City? And (2) who are the most salient stakeholders of household solid waste management in Cimahi City? The research utilised a qualitative method approach. Data collection techniques contained in-depth interviews, non-participant observations and reading documentation. Triangulation was applied to validate the collected data. Even further, data was analysed by the Miles and Huberman model. The result of this research has identified stakeholders with important influence and impacts on

---

V. S. Suherman (✉)
Doctoral Program in Environmental Sciences, Universitas Padjadjaran, Bandung, West Java Province, Indonesia

M.-L. Franco-García
University of Twente, Enschede, The Netherlands
e-mail: m.l.francogarcia@utwente.nl

O. S. Abdoellah · D. Kurniadie · Y. A. Hidayati
Graduate School on Environmental Sciences, Universitas Padjadjaran, Bandung, West Java Province, Indonesia
e-mail: yuli.astuti@unpad.ac.id

M.-L. Franco-García et al. (eds.), *Towards Zero Waste*, Greening of Industry Networks Studies 6, https://doi.org/10.1007/978-3-319-92931-6_5

the household solid waste management in Cimahi City. The identified stakeholders were classified into two categories: (1) formal sector (government, NGOs and private ones) and (2) informal sector (not officially registered waste collectors and recycling entities). The stakeholder salience for household solid waste management depends on their type of activities, which were reducing and handling. In the reducing activities, one of the "definitive" stakeholders affiliates to the Ministry of Environment and Forestry. Whilst for the handling activities, the "definitive" stakeholder comes from the municipality of Cimahi City. Understanding the identity and saliency of stakeholders will help develop household solid waste management strategies with circular economy principles.

Keywords Stakeholders analysis · Stakeholders salience · Household solid waste · Waste management

5.1 Introduction

Local governments are facing some serious challenges to solve solid waste management problems in both developed countries (Ehrlich and Ehrlich 1972) and developing countries (Guerrero et al. 2013; Marshall and Farahbakhsh 2013; Suttibak and Nitivattananon 2008). Those challenges come as a combination of the increase in waste volume, landfill space depletion and problems related to gain a new landfill (Suttibak and Nitivattananon 2008). Furthermore, governments have failed to provide adequate solid waste management services because of their lack of capacities to adopt intensive technologies from developed countries (Okumu and Nyenje 2011). In addition to that, financial, social and legislation problems represent some of the main solid waste management issues in the developing countries (Damanhuri 2010; Ernawati et al. 2012; Henry et al. 2006; Okumu and Nyenje 2011).

Indonesia, as most of the developing countries, presents a complex household solid waste management, particularly in big cities such as Jakarta, Bandung, Yogyakarta, Semarang, Surabaya and Medan (Damanhuri 2010 and Ernawati et al. 2012). The Indonesian government realises the importance of an adequate solid waste management for the environment and quality of life. For that reason, the government issued the Solid Waste Management Law No. 18/2008. This law was created after the Leuwigajah dumpsite disaster on the 21st of February 2005, which killed more than 143 people in that area. According to this law, solid waste management activities consist of reduction and handling. Reducing activities include those related to waste generation limitation, recycle and reuse, whereas handling activities consist of separation, collection, transportation, processing and final processing.

Despite the existing legislations, the solid waste management problems in Indonesia are still unresolved. The large dimension of solid wastes reflects those problems. In 2006, the total amount of solid waste reached 38.5 million tons (MoE 2008), and it increased to 178.85 million tons in 2012 (MoE 2013). Kastaman and

Kramadibrata (2007) stated that the main problem of solid waste management in urban areas of Indonesia was inadequate planning, utilisation and participation to handle and manage solid waste and the constraint of activities to enhance solid waste value economically.

On the one hand, Muthmainnah (2007) emphasised the low level of public awareness to participate in solid waste management as one of the failure causes of solid waste management in Indonesia. On the other hand, production and consumption patterns represent a second failure cause because they affect the volume and characteristics of wastes.

There were some studies to address issues of sustainable solid waste management in Indonesia by using various tactics such as the use of clean technologies (Sulistyorini 2005; Kastaman and Kramadibrata 2007; Damanhuri 2010), social approaches (Muthmainnah 2007; Utami et al. 2008; Budiman, et al. 2013), economic methods (Aye and Widjaya 2005; Sejati 2009) and institutional and legal approaches (Ernawati et al. 2012). Other authors, such as Seadon (2006), stated that the success of solid waste management could be achieved if there were more *change agents* involved along the different phases of the waste management. The change agents come from a different perspective and have the ability to communicate efficiently among different stakeholders. As the result, the identification of crucial stakeholders is important in solid waste management (Seadon 2006). Heidrich et al. in 2009 proposed some of the advantages of including stakeholder analysis in solid waste management such as reduced pollution and disposal cost. These are the result of stakeholders' efforts to redefine priorities and redirect strategies whilst anticipating problems and issues. Stakeholder analysis can be successfully implemented to gain better understanding of stakeholder roles and actions, analysing the driving factors and coordination among stakeholders, as well as the identification of obstacles in communication that affect daily operational or planning strategy (Caniato et al. 2014).

Furthermore, stakeholder analysis provides a basic understanding of important stakeholders' inclination to engage actively in circular economy of waste management. This as part of the argumentation extensively found in literature in relation to the concept of circular economy which represents a convincing strategy to reduce input of raw materials and output of waste in economic and ecological circularity through recycling and reuse (Haas et al. 2015). The implementation of circular economy tenets requires the collaboration of diverse actors within the solid waste management system. Consequently, the involvement of stakeholders is crucial to prevent failure in waste management systems (Caniato et al. 2014). Therefore, it is important to understand the role of stakeholders in household solid waste management. This paper discusses stakeholder identity and salience theory from Mitchell et al. (1997) to address different attributes of stakeholders and the most prominent stakeholders who have a large impact to solve household solid waste.

The first part of this paper will present the research background of household solid waste management problems in most of the developing countries including Indonesia and reasons to choose the stakeholder identity and salience theory from Mitchell et al. (1997). Then, this theory will be elaborated in the second part of this paper. Section 5.3 explains current household solid waste management in Cimahi City. The method of this research is described in Sect. 5.4. Results and discussions are presented in Sect. 5.5. Finally, Sect. 5.6 presents the conclusion of this research.

5.2 Stakeholder Identity and Salience Theory

Stakeholder theory attempts to identify the primary stakeholders that deserve or require managerial attention (Mitchell et al. 1997). There are many definitions of stakeholders, and they often have the same roots as it is stated by Freeman (1984), who defines stakeholders as "any group or individual that affects or is affected by the achievement of the organisation's objectives". Clarkson (1995) defines stakeholders as persons or groups that have or claim ownership, rights or interests in a corporation and its activities, past, present or future. Freeman's definition of stakeholders was one of the broadest definitions in literature, meanwhile Clarkson's definition was the narrowest one (Mitchell et al. 1997). In fact, there are many definitions of stakeholders, which all lay on the spectrum that exists between broadest and narrowest definitions, and these definitions have been used by Mitchell et al. (1997) to form core attributes of his stakeholder identification model. This model has been applied in this paper.

Stakeholder theory is usually used for corporation management purposes (Heidrich et al. 2009), but some studies used this theory for other fields such as project management (Aaltonen et al. 2008; Achtercamp and Vos 2008), social management (Crane and Ruebottom 2012), agriculture (Hoppe and Sanders 2014), natural environment (Reed et al. 2009) and waste management (Srivastava et al. 2005; Heidrich et al. 2009; Caniato et al. 2014). Although there are some studies about waste management using stakeholder theory, there is little evidence of research carried out to analyse all of the stakeholders engaged in the household solid waste management, particularly in Indonesia. Meanwhile, stakeholders play an important role for sustainable solid waste management (Joseph 2006; Heidrich et al. 2009; Zurbrügg et al. 2012). Hence, it is important to have knowledge of relevant stakeholders and how they might be managed appropriately in the waste management process (Heidrich et al. 2009). Even though it can be difficult to understand the composition and working of different types of stakeholder networks, and the ways their effectiveness can be maximised (Caniato et al. 2015).

Guerrero et al. (2013) described stakeholders as one of those factors which strongly influence the waste management system and are essential for the clear understanding of stakeholder's responsibilities. By using the stakeholder identity and salience theory developed by Mitchell et al. (1997), three attributes (*power, legitimacy, and urgency*) which shape the stakeholder's typology were applied to classify the stakeholders of the household solid waste management in Cimahi City. As indicated by Mitchell, "power" can be defined as the ability possessed by stakeholders to impose their will on a certain relationship through normative[1], coercive[2] and utilitarian[3] power (Etzioni 1964). The "legitimacy" term was described as the generalised perception or assumption that the actions of an entity are desirable, proper or appropriate within the context of the social system through individual, organisational and societal action. In the case of "urgency", it is defined as the degree to which a stakeholder claims to have time sensitivity or criticality.

The stakeholder identity and salience theory produced a comprehensive typology of stakeholders and a dynamic model which asserts that stakeholder status is not fixed. It can be changed based on determination of decision-makers (Magness 2007). In accordance to Mitchell et al. (1997), there are eight types of stakeholders, out of which three types of stakeholders (dormant[4], discretionary[5], demanding[6]) have only one attribute, three of them (dominant[7], dangerous[8], dependent[9]) have two attributes and one of these stakeholders (definitive stakeholders[10]) has all **three attributes**. Stakeholders are said to be salient if they have power, legitimacy and urgency attributes at the same time. Stakeholder salience is defined as "the degree to which managers give priority to competing stakeholder claims" (Mitchell et al. 1997). In addition, those stakeholders who have a relationship with the company but do not possess power, legitimacy and urgency are referred as non-stakeholders (type 8).

Later on, Drisscoll and Starik (2004) were widening the categorisation of stakeholders from human to non-human. They added proximity as an attribute that was connected to the near-far, the short-long term and the actual-potential dimensions.

---

[1] Control based on application of physical means is ascribed as *coercive power* (Etzioni 1964).

[2] The use of material means for controlling purpose constitutes *utilitarian power* (Etzioni 1964).

[3] The use of symbols for control purposes is referred to as *normative*, *normative-social* or social power (Etzioni 1964).

[4] Dormant stakeholders possess *power as attribute* to impose their will on the firm, but by not having a legitimate relationship or an urgent claim, their power remains unused (Mitchell et al. 1997).

[5] Discretionary stakeholders possess the *attribute of legitimacy*, but they have no power to influence the firm and urgent claims (Mitchell et al. 1997).

[6] Demanding stakeholders, those with urgent claims (*urgency attribute*) but having neither power nor legitimacy, are the "mosquitoes buzzing in the ears" of managers: irksome but not dangerous, bothersome but not warranting more than passing management attention, if any at all (Mitchell et al. 1997).

[7] Dominant stakeholders: in the situation where stakeholders are both *powerfull and legitimate*, their influence in the firm is assured, since by possessing power with legitimacy, they form the *dominant coalition* in the enterprise (Mitchell et al. 1997).

[8] Dangerous stakeholders: where *urgency and power* characterise a stakeholder who lacks legitimacy, that stakeholder will be coercive and possibly violent, making the stakeholder "dangerous", literally, to the firm (Mitchell et al. 1997).

[9] Dependent stakeholders, who lack power but who have urgent legitimate claims as dependent (*Urgency and legitimate attributes*), because these stakeholders depend upon others (other stakeholders or the firm's managers) for the power necessary to carry out their will (Mitchell et al. 1997)

[10] Definitive stakeholders: when dominant stakeholder's claim is urgent, managers have a clear and immediate mandate to attend to and give priority to that stakeholder's claim. The most common occurrence is likely to be the movement of dominant stakeholder into the "definitive" category (Mitchell et al. 1997).

The "proximity" attribute made the natural environment a primary and primordial stakeholder of corporation. They argued that the stakeholder theory from Mitchell et al. (1997) is still attached to a social paradigm that prioritises economic and political reasons. Therefore, it needed to be enlarged to admit connections between business organisations and ecological systems. Meanwhile, Heidrich et al. (2009) suggested that the term of salience refers to being important. In addition to this, Neville et al. (2011) tried to refine and redefine the stakeholder identification and salience theory with a new category which excluded the urgency attribute in identification of stakeholders. Nevertheless, all different approaches to Mitchell's theory concluded that the moral legitimacy is the most important thing that applies to stakeholder salience. More importantly, they stated that the salience of a stakeholder would vary according to his/her level of attributes, as well as accuracy in stakeholder identification and stakeholder salience assessment (Neville et al. 2011). Consequently, for this paper, stakeholder identity and salience theory is used as it was developed and defined by Mitchell et al. (1997), in order to provide further understanding about authority and responsibility that should be taken by certain stakeholders in the household solid waste management.

Although Mitchell et al. (1997) had developed eight types of stakeholders. Parent and Deephouse (2007) stated that the most significant in their study were the dominant, dormant or definitive types. The other five types have rarely been used. The most influential attribute to salience was power, followed by urgency and legitimacy. Eesley and Lenox (2006) stated that in order to maintain the power, stakeholders need access to resources that are referred by Etzioni (1964) as utilitarian power. Parent and Deephouse (2007) have proposed that there are only three types of stakeholders that are most commonly present in a household solid waste management system. However, this paper focuses more on the identification of roles and attributes of those stakeholders, which have been identified by Mitchell et al. (1997). We believe that the success of household solid waste management depends on stakeholder relationships. This theory can identify the salient stakeholders that have the most influence, to improve the household waste management system, looking in particular to the case in hand: Cimahi City, West Java Province, Indonesia.

5.3 Household Solid Waste Management in Cimahi City

Indonesia is the fourth most populous country in the world with 248,818,100 inhabitants in the year 2013 (BPS-Statistics Indonesia 2014). 57.5% of the population lives on the island of Java. West Java Province is the most populous province in Indonesia having 43,053,700 inhabitants, which correspond to 18.11% of the total population in Indonesia (BPS-Statistics Indonesia 2014). As one of the developing countries with a dense population, Indonesia has a complex solid waste management situation, particularly in big cities such as Jakarta, Bandung, Yogyakarta, Semarang, Surabaya and Medan (Damanhuri 2010 and Ernawati et al. 2012).

Cimahi is the only city in Indonesia that comprises three sub-districts and ranks second in West Java Province as the most densely populated city with 13,859 inhabitants/km^2 of urban density (BPS of West Java Province 2014). Cimahi is 40.25 hectares wide; 40% of the land in Cimahi belongs to the military and cannot be utilised by others institutions, communities or corporations. Cimahi has more than 600,000 inhabitants and 130 formal industries that have to share the 60% of the remaining land. Industrial activities are dominated by textile, foods and leather processing industries. Consequently, Cimahi is facing serious environmental problems including household solid waste generation, particularly after the Leuwigajah dumpsite disaster on 21 February 2005. This disaster occurred after a large slide of waste hit villages nearby and buried 71 houses and killed 143 people. It happened because of poor solid waste management. Since that disaster, Cimahi does not have a landfill to manage its own waste any more. To date, Cimahi has been delivering its waste to Sarimukti dumpsite, West Bandung Regency. The distance of the dumpsite is approximately 33 km from Cimahi, and this dumpsite can be used only temporarily. The land belongs to Perum Perhutani (a stated-owned company to manage Indonesian forestry). The contract with Perum Perhutani will end in 2017, and the main reason for this is the limited capacity of the Sarimukti dumpsite.

Therefore, the dumpsite area will move to Legok Nangka, Garut Regency, within approximately 45 km from Cimahi. As a result, the cost of waste management in Cimahi will increase because the provincial government of West Java has appointed the Legok Nangka dumpsite as regional landfill for four regencies and cities in West Java (Cimahi City, Bandung City, Bandung Regency and Bandung Barat Regency). In addition, the West Java government announced Legok Nangka as the last possible land to become dumpsite area. For that reason, the West Java Province coordinates with the local government in Bandung City, Bandung Regency, Bandung Barat Regency and Cimahi City for waste reduction and handling. As for the waste-reducing activities, the Cimahi government enforces a 3R approach (Reduce, Reuse and Recycle) which still has not shown significant results as can be observed from data of waste generation in Cimahi City. In 2007, the population in Cimahi was 536,743 inhabitants and the amount of waste was 1381.28 m^3/day (SLHD 2008). In 2011, the population in Cimahi had risen to 612,168 inhabitants, whilst the amount of waste was 1407 m^3/day (SLHD 2011). Meanwhile, in waste-handling activities, Cimahi is still using a "linear" collect-transport-dispose system, and it still relies on domestic waste transport fleets. This system is known as the conventional waste management system (Kastaman and Kramadibrata 2007). According to data from the environmental ministry (MoE 2013), more than 50% of solid waste came from household wastes, and it could not be handled properly. The solid waste volume in Indonesia was estimated to be 1 million cubic metre per day, but only 42% of it was transported and processed properly.

Figure 5.1 shows the household solid waste management system in almost all districts and cities in Indonesia, including Cimahi City. The household solid waste handling can be done in three different ways. The first way is that the household solid waste that has been collected is transported by open waggons to temporary dumpsites and transported by trucks to be dumped directly into the final dumpsite.

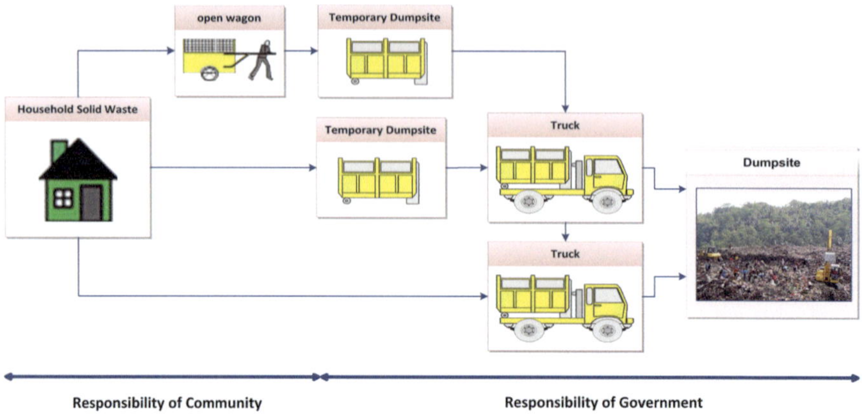

Fig. 5.1 Household solid waste management in most of Indonesian regencies/cities. (Adapted from Cimahi City Government report)

Secondly, the household solid waste is directly collected in the temporary dumpsites and transported by trucks into the final dumpsite area. The third way is that the household solid waste that has been collected is directly transported by trucks to be dumped into the final dumpsite without passing through temporary dumpsites. Transporting household solid waste from households to temporary dumpsites is the responsibility of communities, whilst transporting household solid waste from temporary dumpsites into the final dumpsite is the responsibility of the districts/cities. The procedure of household solid waste transport has been regulated by the Ministry of Interior No. 33/2010, on guidelines for solid waste management and regional regulation of Cimahi City No.16/2011 about waste management.

Prior to 2001, solid waste management responsibilities mostly were distributed among some departmental and central governments. After decentralisation, local government obtained a greater role and responsibility than the central government in household solid waste management. According to Bruce and Storey (2010), waste management in Indonesia as it has been decentralised provides an opportunity to resolve the problem locally. But, it also opens up the problem of limited capacity, resources and inability to perform coordination among districts/cities and other institutions. The Law No. 18/2008 on waste management has granted greater autonomy and scope to local and regional governments, private entities and communities to manage their own waste systems. Although to date, reviews on the division of responsibility emphasise more the confusion and conflict on the responsibilities and resources (Bruce and Storey 2010). The problem that usually arises in the relationship between the stakeholders includes issues about the division of authority between local, regional and national levels, government and society, government and business or the government and the informal sector (scavengers, waste dealers).

5.4 Methodology

This research utilised a qualitative method approach and a literature review. The qualitative method is used to explore and to gain good understanding of household solid waste management problems, as mentioned by Creswell (2002). Data collection was done by in-depth interviews, non-participant's observations and reading documentation. The informants were selected by a purposive sampling technique. The aim of this technique is to describe the heterogeneity of communities, in this case the stakeholders, resulting in more variation of possibilities in developing conclusions based on different settings or individual opinions. In addition, various informational sources, i.e. official documents, books and peer-reviewed journals, were consulted.

In order to facilitate the description of the household solid waste management stakeholders in Cimahi, two categories were created, those who belonged to the formal sector and those belonging to the informal sector. According to Bruce and Storey (2010), the formal sectors are the government and listed companies that usually serve as a regulator and official provider of waste management services. The informal sector usually is dominated and motivated by market-driven and economic factors for performing the necessary resource recovery and collection services. For the purpose of this research, stakeholders from the formal sector consist of governmental groups, NGOs and the private sector (corporation and recycling companies with permission from the government), whilst the informal sector of stakeholders corresponds to public and waste management small businesses that do not have official registration.

The stakeholder selection in previous studies usually depended on the choice of researchers. Therefore, to ensure that all stakeholders are identified, this study followed the method performed by Parent and Deephouse (2007) that allows managers to evaluate overall stakeholder issues. Parent and Deephouse (2007) found that the role and position of the interviewee in the organisational structure affect the identification of stakeholders. Top managers typically provided a wider range of stakeholders. Consequently, various stakeholders have been interviewed.

The data were analysed by using the Miles and Huberman model (1984). According to it, data analysis included (1) data reduction, (2) data display and (3) conclusion drawing/verification. Data reduction can be defined as the process of selecting, focusing, simplifying, abstracting and transforming the data into transcription or written field notes. As the data collection process proceeds, the data reduction continues as well. After the fieldwork, the data collection keeps proceeding until a final report is formed. Data display is referred as the organisation and compression of information, which allows the drawing of conclusions and possibly proposing some specific actions. The conclusions are verified as the analyst proceeds. In this study, data analysis began when the interviews were started and continued after they were done. During interviews, informants' answers were analysed and crosschecked. This process continued until there was no additional data included.

This research was conducted in Cimahi City, West Java Province, Indonesia. The location was chosen because of its complex environmental problems.

5.5 Relevant Findings and Discussion

The Law No. 18/2008 on waste management regulates roles, obligations and rights of each stakeholder related to waste management, namely, the central government, the provincial government, the local government, the private sector and communities. These stakeholders are enlisted in Tables 5.1 and 5.2. This law also states that the waste management activities consist of waste reducing and handling. Based on Law No. 33/2012 about regional authority, the solid waste problem affairs are governed by the environmental and public departments of all government levels (local, central and provincial). In fact, solid waste affairs are a matter of the joint collaboration among the Ministry of Interior, the Ministry of Trade and Industry, the Ministry of Environment and Forestry and the Ministry of Public Works at the central level. Each of the mentioned ministries issues regulation on household solid waste management, and those are expected to be implemented at provincial and local level. At the provincial level, there are the Regional Environmental Protection Board (BPLHD) and the Housing and Settlement Department (Diskimrum). At the local level, the involved stakeholders can be different depending on the district and the city itself. In Cimahi, the involved stakeholders are the Cleanliness and Landscape Department (DKP), the Planning Agency (Bappeda) and the Environmental Agency of Cimahi City (KLH). In addition, the private sectors, NGOs and informal sectors play an important role in household solid waste management.

The involvement of various stakeholders causes problems related to the division of authority among them. Therefore, it is important to identify the involved stakeholders in reducing and handling waste management activities in both the formal sectors and the informal sectors. Furthermore, this study is focused on salient stakeholders in order to seek the promoters and coordinating actors who can improve the household solid waste management problems in Cimahi. The early stage of this study indicated that the attributes of stakeholders of household solid waste management vary depending on the two main waste management activities stated in the Indonesian law: reducing and handling.

Table 5.1 shows the identity and salience of involved stakeholders in waste-reducing activities. The definitive stakeholder in this activity based on Law No.18/2008 should be the Ministry of Environment and Forestry, whilst other governmental institutions qualify as dominant stakeholders. Seadon (2006) proposed that the overall responsibility for solid waste management processes should lie on the government since they are the prime change enablers. However, in reality, most governments are (just) dominant stakeholders. This is because there are conflicts of responsibility among local, provincial and central governments. Based on interviews, each of the governmental institutions pointed to another institution as the one who has greater responsibilities! For instance, local government expected instructions and support (financial, technical and also material) from provincial government to start waste-reducing programmes. In contrary, the provincial government indicated that the central government (the Ministry of Environment and Forestry) is the most responsible stakeholder. But the central government mentioned

Table 5.1 Stakeholders identity and salience for reducing activities in household solid waste management (formal sectors)

No	Stakeholders	Roles	Power (dormant)			Legitimacy (discretionary)			Urgency (demanding)		Power + legitimacy (dominant)	Power + urgency (dangerous)	Legitimacy + urgency (dependent)	Power + legitimacy + urgency (definitive)
			C[a]	U[b]	N[c]	I[d]	Or[e]	S[f]	TS[g]	Cr[h]				
	A. Formal sectors													
1.	**Government**													
	National government authority													
	1.1. Determine national policies and strategies in solid waste management (Law No. 18/2008; Bruce and Storey 2010)													
	1.2. Determine norms, standards, procedures and criteria in solid waste management (Law No. 18/2008; Bruce and Storey 2010)													
	1.3. Facilitating and developing interregional cooperation, partnership and networking in solid waste management (Law No. 18/2008; Bruce and Storey 2010)													
	1.4. Provide coordination, coaching and monitoring of local government performance in solid waste management (Law No. 18/2008; Bruce and Storey 2010)													
	1.5. Determine policies for interregional disputes and solutions in solid waste management (Law No. 18/2008; Bruce and Storey 2010)													
	Ministry of Environment and Forestry	Officially												
		Point 1.1, 1.3, 1.4 and 1.5			√		√			√				√
		Practically												
		Point 1.1, 1.3, 1.4 and 1.5					√				√			
	Ministry of Public Works	Officially												
		Point 1.2 and 1.4			√		√				√			
		Practically												
		Point 1.2 and 1.4		√			√				√			

(continued)

Table 5.1 (continued)

No	Stakeholders	Roles	Power (dormant)			Legitimacy (discretionary)			Urgency (demanding)		Power + legitimacy (dominant)	Power + urgency (dangerous)	Legitimacy + urgency (dependent)	Power + legitimacy + urgency (definitive)
			C[a]	U[b]	N[c]	I[d]	Or[e]	S[f]	TS[g]	Cr[h]				
	Ministry of Trade and Industry	Officially												
		Point 1.2			√		√				√			
		Practically												
		Point 1.2		√			√				√			
	Provincial government authority													
	1.1. Determine provincial policies and strategies in line with national government policies (Law No. 18/2008; Bruce and Storey 2010)													
	1.2. Facilitate intra-provincial cooperation, partnership and networking (Law No. 18/2008; Bruce and Storey 2010)													
	1.3. Monitor and support local district and municipality governments in waste management (Law No. 18/2008; Bruce and Storey 2010)													
	1.4. Facilitate intra-provincial dispute solution (Law No. 18/2008; Bruce and Storey 2010)													
	Regional Environmental Protection Board of West Java Province	Officially												
		Point 1.1, 1.2, 1.3 and 1.4			√		√				√			
		Practically												
		Point 1.1, 1.2, 1.3 and 1.4			√		√				√			
	Housing and Sett lement Department of West Java Province	Officially												
		Point 1.2, 1.3 and 1.4			√		√				√			
		Practically												
		Point 1.2, 1.3 and 1.4		√			√				√			

	Local government authority													
	1.1. Apply and enforce government waste management strategies and policies (Law No. 18/2008; Bruce and Storey 2010)													
	1.2. Apply and enforce government waste management norms, standards, procedures and criteria (Law No. 18/2008; Bruce and Storey 2010)													
	1.3. Monitor external third party waste management (Law No. 18/2008; Bruce and Storey 2010)													
	1.4. Maintain official dumps, landfills, collection sites and other disposal/treatment facilities (Law No. 18/2008; Bruce and Storey 2010)													
	Cleanliness and Landscape Department of Cimahi City	Officially												
		Point 1.1, 1.2, 1.3 and 1.4			√		√				√			
		Practically												
		Point 1.1, 1.2, 1.3 and 1.4					√		√				√	
	Planning Agency of Cimahi City	Officially												
		Point 1.3			√		√				√			
		Practically												
		Point 1.1 and 1.3		√			√				√			
	Environmental Agency of Cimahi City	Officially												
		Point 1.1 and 1.3			√		√				√			
		Practically												
		Point 1.3					√							
2.	**Non-governmental organisations (NGOs)—national level**													
	2.1. Any person in household solid waste management shall reduce and handle waste in environmentally friendly manner (Law No. 18/2008)													
		Officially												
		Point 2.1			√		√				√			
		Practically												
		Point 2.1		√			√				√			

(continued)

Table 5.1 (continued)

No	Stakeholders	Roles	Power (dormant)			Legitimacy (discretionary)			Urgency (demanding)		Power + legitimacy (dominant)	Power + urgency (dangerous)	Legitimacy + urgency (dependent)	Power + legitimacy + urgency (definitive)
			C[a]	U[b]	N[c]	I[d]	Or[e]	S[f]	TS[g]	Cr[h]				
3.	**Private sector**													
	3.1. Producer required to manage the packaging and/or product which cannot or is difficult to decompose by natural processes (Law No. 18/2008)													
	3.2. Any person who carries on business in waste management are required to have a permit from the head of the region (Law No. 18/2008)													
	Corporation	Officially												
		Point 3.1					√		√				√	
		Practically												
		Point 3.1					√							
	Formal recycler industries	Officially												
		Point 3.2					√			√			√	
		Practically												
		Point 3.2		√			√				√			
4.	**Communities**													
	4.1. Reduce and handle waste in environmentally friendly manner (Law No. 18/2008)													
		Officially												
		Point 4.1		√				√			√			
		Practically												
		Point 4.1						√						

[a]Coersive
[b]Utilitarian
[c]Normative
[d]Individual
[e]Organisational
[f]Social
[g]Time sensivity
[h]Critically

that provincial and local governments are the main stakeholders to carry out these activities. Even though, they admitted the obligation of the ministry to provide law and enforcement instruments for reducing waste generation from goods manufacturing companies, but they have limitations to enforce the law themselves due to political, financial, technological and social problems.

Furthermore, due to the DKP mandate, it should be a dominant stakeholder, but it is now becoming a dependent stakeholder due to its financial, technological and human resource constraints. DKP needs to rely on other stakeholders, such as the central government and the provinces. KLH is supposed to be a dominant stakeholder, as well, though at present this institution has become a discretionary stakeholder because it has legitimacy but lack of power and urgency. In addition, some of the KLH functions can be transferred to the Planning Agency (Bappeda). This happens because Bappeda has a utilitarian power (see Section 5.2) that according to Parent and Deephouse (2007) has a greater effect on the salience.

The NGOs that are involved in waste-reducing activities are dominant stakeholders because they have power and legitimacy. Officially, NGOs have normative power, but practically, they have utilitarian power due to material support. Private sector companies are divided into corporations and formal recycling companies that should be a dependent stakeholder. This is because, and to some extent, they still require assistance from the government to implement the waste reduction methods. However, at present the corporations are discretionary stakeholders because they only have legitimacy. They do not use their power and exclude their urgency in waste management. The formal recycling companies are a dominant stakeholder because they have power to regulate and set the price of recycled goods which can be used for their companies. The community members, as established by the Law No. 18/2008, should be the dominant stakeholders in reduction of household solid waste management because they represent the major source of the waste. Their consumption patterns cause the solid waste generation (Muthmainnah 2007; Falasca-Zamponi 2011). Nevertheless, the community is categorised as a discretionary stakeholder because of the lack of participation and public awareness in the management of household waste (DKP, Cimahi 2014).

Whilst Table 5.1 showed the stakeholders in waste reduction, on the other hand, Table 5.2 displays the identity and salience of involved stakeholders in waste-handling activities. The involved stakeholders in these activities are both formal and informal sectors. DKP should be considered as a definitive stakeholder, but now, DKP is a dependent stakeholder. It is because DKP still relies on other stakeholders including Bappeda, which has the authority to establish the work, to develop the programmes and to allocate budgets for DKP. Bappeda is categorised as a dominant stakeholder, but practically, it becomes a definitive stakeholder due to its authority to set development planning programmes for other departments. They have capacity to insert or remove the programmes that will be implemented. Mitchell et al. (1997) mentioned that a stakeholder exhibiting both power and legitimacy could be moved from the category of dominant stakeholder to definitive stakeholder when such stakeholder's claim is urgent. Managers have a clear and immediate mandate to attend to, and they give priority to that stakeholder's claims.

Table 5.2 Stakeholders identity and salience for handling activities in household solid waste management (formal and informal sectors)

No	Stakeholders	Roles	Power (Dormant)			Legitimacy (Discretionary)			Urgency (Demanding)		Power + Legitimacy (Dominant)	Power + Urgency (Dangerous)	Legitimacy + Urgency (Dependent)	Power + Legitimacy + Urgency (Definitive)
			C	U	N	I	Or	S	TS	Cr				
A. Formal Sectors														
1.	**Government**													
	National Government Authority													
	1.1 Determine national policies and strategies in solid waste management (Law No. 18/2008; Bruce and Storey, 2010)													
	1.2 Determine norms, standards, procedures and criteria in solid waste management (Law No. 18/2008; Bruce and Storey, 2010)													
	1.3 Facilitating and developing inter-regional cooperation, partnership, and networking in solid waste management (Law No. 18/2008; Bruce and Storey, 2010)													
	1.4 Provide coordination, coaching and monitoring of local government performance in solid waste management (Law No. 18/2008; Bruce and Storey, 2010)													
	1.5 Determine policies for inter-regional disputes solutions in solid waste management (Law No. 18/2008; Bruce and Storey, 2010)													
	Ministry of Environmental and Forestry	Officially												
		Point 1.1, 1.3, 1.4 and 1.5			√		√				√			
		Practically												
		Point 1.1, 1.3, 1.4 and 1.5					√							
	Ministry of Public Work	Officially												
		Point 1.2 and 1.4			√		√				√			
		Practically												
		Point 1.2 and 1.4		√			√				√			
	Provincial Government Authority													
	1.1 Determine provincial policies and strategies in line with national government policies (Law No. 18/2008; Bruce and Storey, 2010)													
	1.2 Facilitate intra-provincial cooperation, partnership and networking (Law No. 18/2008; Bruce and Storey, 2010)													
	1.3 Monitor and support local district and municipality governments in waste management (Law No. 18/2008; Bruce and Storey, 2010)													
	1.4 Facilitate intra-provincial dispute solution													

	Regional Environmental Protection Board of West Java Province	Officially												
		Point 1.1, 1.2, 1.3 and 1.4			√		√				√			
		Practically												
		Point 1.1, 1.2, 1.3 and 1.4			√		√				√			
	Housing and Settlement Department of West Java Province	Officially												
		Point 1.2, 1.3 and 1.4		√			√				√			
		Practically												
		Point 1.2, 1.3 and 1.4		√			√				√			
	Local Government Authority													
	1.1 Apply and enforce government waste management strategies and policies (Law No. 18/2008; Bruce and Storey, 2010)													
	1.2 Apply and enforce government waste management norms, standards, procedures and criteria (Law No. 18/2008; Bruce and Storey, 2010)													
	1.3 Monitor external third party waste management (Law No. 18/2008; Bruce and Storey, 2010)													
	1.4 Maintain official dumps, landfills, collection sites and other disposal/treatment facilities (Law No. 18/2008; Bruce and Storey, 2010)													
	Cleanliness and Landscape Department of Cimahi City	Officially												
		Point 1.1, 1.2, 1.3 and 1.4			√		√		√					√
		Practically												
		Point 1.1, 1.2, 1.3 and 1.4					√		√				√	
	Planning Agency of Cimahi City	Officially												
		Point 1. 3			√		√				√			
		Practically												
		Point 1.1 and 1.3		√			√			√				√
	Environmental Agency of Cimahi City	Officially												
		Point 1.1 and 1.3			√		√				√			
		Practically												
		Point 1.3					√							

(continued)

Table 5.2 (continued)

No	Stakeholders	Roles	Power (Dormant)			Legitimacy (Discretionary)			Urgency (Demanding)		Power + Legitimacy (Dominant)	Power + Urgency (Dangerous)	Legitimacy + Urgency (Dependent)	Power + Legitimacy + Urgency (Definitive)
			C	U	N	I	Or	S	TS	Cr				
2.	**Non-Governmental Organizations (NGOs)-Local Level**													
	2.1. Any person in household solid waste management shall reduce and handle waste in environmentally friendly manner (Law No. 18/2008)													
		Officially												
		Point 2.1			√		√				√			
		Practically												
		Point 2.1		√			√				√			
3.	**Communities**													
	3.1. Reduce and handle waste in environmentally friendly manner (Law No. 18/2008)													
		Officially												
		Point 4.1						√	√				√	
		Practically												
		Point 4.1						√						
B. Informal Sectors														
1.	**City's waste manager**													
		Practically												
		Waste handling operators												
		Waste recyclable collectors								√				

Bappeda has used its authority to claim the urgency and has been categorised as a definitive stakeholder. The NGOs and the communities have both some attributes that help them to be involved in the reduction and handling of solid waste by advocating government policy and supporting government programmes to minimise waste generation.

The informal sector consists of a city's waste manager, informal waste collectors and informal recycling companies. The city's waste manager represents a demanding stakeholder because they only have the urgency attribute when they are unable or unwilling to acquire either the power or the legitimacy necessary to move their claim into a more salient status (Mitchell et al. 1997). They are assigned to collect solid waste from households, streets and shopping and trade centres which is then dumped to the temporary dumpsite. Before the disposal of household solid waste to the temporary dumpsite, the marketable goods are taken out which helps in the generation of additional income. This act is not considered as an illegal one in Indonesia since it is beneficial in a way that it helps in the reduction of the waste, which is then sent to the final dumpsite (Damanhuri 2009).

Informal waste collectors are also categorised as a demanding stakeholder because they do not have power and legitimacy, but their presence in the waste-handling activities is crucial. The informal recycling industry is assessed as dangerous stakeholder because it has a utilitarian power and urgency in most of the cases. They are a dangerous stakeholder because they can obstruct government programmes to reduce solid waste generation. For example, a restriction by government to use plastic bags will threaten the continuity of their company because it reduces input of recycled plastics that normally are used as raw material. As for the informal companies, they also have the ability to determine the price of incoming goods. These types of small companies are categorised as informal because they do not register to the Industrial Chamber and do not pay taxes. The owners of informal companies believe that becoming a part of the formal sector is not easy and it brings inconvenience. Some of the barriers to registration are the procedural difficulties, costs and requirements to fulfil the registration processes (Bruce and Storey 2010).

Based on the above discussion, it can be observed that stakeholders, who are involved in the household solid waste management in Cimahi, are different regarding the waste management activities: reduction and handling. This is because for the managing of solid waste, the government has created legislation that has divided the managerial authority among central, provincial and local governments. This has directly affected the government's performance in solid waste management, especially by lowering the priority in handling waste due to financial problems and by the occurrence of locality ego-centrism, which creates difficulties for local governments to operate landfills that are situated in the areas outside their jurisdiction, as it has mentioned by Damanhuri (2008). In addition to it, there are conflicts among stakeholders on using their authorities or as it was called by Bruce and Storey (2010) "implementation confusions" among stakeholders. Furthermore, there is still a huge gap between human resource competencies and requirement competencies in Cimahi, which was a similar result compared to the study carried out by Herayani (2011).

Furthermore, the results of this study show that a majority of the stakeholders are dominant, dependent and definitive. This contradicts the findings of Parent and Deephouse (see Sect. 5.2). This is because stakeholder attributes are not fixed and they can be changed by particular entity or determination from decision-makers (compare Mitchell et al. 1997; Magness 2007). Besides that, the most influential factor to cause saliency in household solid waste management in Cimahi is utilitarian power. According to Etzioni (1964), stakeholders need access to resources in order to maintain their power. Therefore, NGOs and Bappeda represent those stakeholders that have used utilitarian power to claim their attributes. NGOs have material supports as utilitarian power, whilst Bappeda has utilitarian power to set goals in development planning programmes for other departments.

The determination of stakeholder's identity and salience is very important when there are multi-stakeholders' activities. This classification can shed light on those stakeholders who can have strong influence over the processes needed for the solution of a problem, the waste management in Cimahi, in this particular case. Even further, if the efforts of one of the stakeholders fail along the process, its impact might produce different damage levels (Clarkson 1995). After analysis of the current stakeholder's constellation of household waste management in Cimahi, some of the highlighted results imply that intervention of the Ministry of Environment and Forestry is very important to boost waste reduction activities. They should become a salient stakeholder who can be able to integrate all of stakeholders that are involved in household solid waste management, as well as to govern and define their roles to minimise conflicts among stakeholders. On the other hand, for the solid waste-handling activities, the intervention of DKP and community participation can improve the waste management system in Cimahi. DKP can provide an adequate technology to handle household solid waste generations, whilst the community can pay retribution as a passive participant to support government financially or by participating actively to reduce, reuse and recycle household solid waste.

However, household solid waste-handling cannot be seen apart from other economic activities. As long as production and consumption continue without closed cycles, solid wastes generation will still exist. Therefore, it is necessary to look for some ways to solve solid waste generation problem, one is by circulating the solid wastes before they are mixed up. Reducing and handling activities are part of material flows which after disposal are available to circulate (reuse or recycle) within the socioeconomic system (Haas et al. 2015). Scheepens et al. (2016) said that the introduction of the circularity in complex systems in a truly circular economy can take several years and that its promotion and coordination rely on the government role who needs to be a reliable stakeholder in the long term. Consequently, it is important to know stakeholders' identity and salience to boost household solid waste management.

5.6 Conclusions

The analysis of stakeholders' identity and salience helps in pointing out the features of those stakeholders who can play an important role to solve household solid waste management problems in Indonesia, particularly in Cimahi City, West Java Province. The stakeholders of the waste management system in Cimahi are identified and divided based on two activities, reduction and handling. The reduction activities are carried out by stakeholders who belong to the formal sector, whilst in handling activities, the stakeholders come from both the formal and the informal sectors. This research was carried out under the assumption that conflicts in household solid waste management can be solved with the intervention of salient stakeholders. It can be concluded that in reducing activities, the involved stakeholders are representatives of governments, corporations, formal recycling industries, NGOs and communities. Even further, according to Mitchell's theory, the Ministry of Environment and Forestry can be qualified as the salient stakeholder. In handling activities, involved stakeholders correspond to formal sector groups, which are governments, NGOs and communities. Whilst for informal sector groups, the city's waste manager, informal waste collectors and informal recycling companies are identified as stakeholders who are involved in household solid waste management. The most salient stakeholder in this activity is DKP. The stakeholder's attributes are mostly dominant, dependent and definitive types, and utilitarian power is the most influential to salience.

The stakeholder identity and salience theory has potency to understand managerial-stakeholder relationships and the identification of most prominent stakeholders in household solid waste management. Since solid waste management has an additional economic value in material circularity, therefore describing stakeholder activities is required for the transition towards sustainable business models in circular economy, and the government should be a reliable stakeholder (Scheepens et al. 2016). Moreover, strong government policies to mainstream circular business are crucial to reap the benefits of circular economy. However, this paper only provides stakeholder identity and salience that can be used to boost waste management in line with the circular economy principles. Hence, further research is foreseen to be carried out to elaborate more on the stakeholder identity and salience under circular economy principles.

In Indonesia, the stakeholder identity and salience for household solid waste management can be different in each city and district. This is because decentralisation legislation has allowed local governments to govern their own cities and districts. However, based on Law No. 33/2012 about regional authority, the affairs of the solid waste management problems are governed by the environmental and public works departments. Hence, the Ministry of Environment and Forestry should be a salient stakeholder to integrate all of stakeholders from local to national level, as well as to govern and define their roles. Consequently, conflicts of authority among stakeholders can be minimised, and household solid waste management in Indonesia can be improved.

References

Aaltonen K, Jaakko K, Tuomas O (2008) Stakeholder salience in global project. Int J Proj Manag 26:509–516

Achtercamp MC, Vos JFJ (2008) Investigating the use of the stakeholder notion in project management literature, a meta-analysis. Int J Proj Manag 26:749–757

Aye L, Widjaya ER (2005) Environmental and economic analyses of waste disposal options for traditional markets in Indonesia. Waste Manag J 26(10):1180–1191

BPS of West Java Province (2014) Jawa barat in figures 2014, Bandung, p 558

BPS-Statistics Indonesia (2014) Statistical yearbook of Indonesia 2014, Jakarta, p 676

Budiman AR, Saam Z, Thamrin (2013) Participation and perception of communities in attempt to maintain, manage environment and defend clean city award. Journal of Ilmu Lingkungan 7(2):103–113

Bruce A, Storey D (2010) Networks of waste: informal economic systems and sustainability in Bali, *Indonesia*. Local Economy Journal 25(3):176–189

Caniato M, Vaccari M, Visvanathan C, Zurbrügg C (2014) Using social network and stakeholder analysis to help evaluate infectious waste management: a step towards a holistic assessment. Waste Management Journal 34:938–951

Caniato M, Tudor T, Vaccari M (2015) Understanding the perceptions, roles and interactions of stakeholder networks managing health-care waste: a case study of the Gaza strip. Waste Management Journal 35:255–264

Clarkson MBE (1995) A stakeholder framework for analyzing and evaluating corporate social performance. Acad Manag Rev 20(1):92–117

Crane A, Ruebottom (2012) Stakeholder theory and social identity: rethinking stakeholder identification. Business Ethics Journal 102:77–87

Creswell JW (2002) Research design: qualitative, quantitative and mixed method approaches. Sage Publications, Thousand Oaks/London/New Delhi, p 246

Damanhuri E (2008) A future prospect of municipal solid waste management in Indonesia. Keynote lecture to the 5th Asian-Pacific Landfill symposium in Sapporo, Japan, October 22(Wed)–24(Fri), 2008

Damanhuri E (2009) Informal collectors of recyclable waste and used goods in Indonesia. in 3R policies for Southeast and East Asia. ERIA Research Project Report 2009, No. 10. p 318

Damanhuri E (2010) Solid and hazardous waste management in Indonesia. A paper. Bandung

DKP Cimahi City (2014) Waste management study of Cimahi City 2014. Cimahi

Drisscoll C, Starik M (2004) The primordial stakeholder: advancing the conceptual consideration of stakeholder status for the natural environment. J Bus Ethics 49:55–73

Ernawati D, Budiastuti S, Masykuri M (2012) Analisis komposisi, jumlah dan pengembangan strategi pengelolaan sampah di wilayah pemerintah Kota Semarang berbasis analisis SWOT. Ekosains Journal IV(2):13–22

Ehrlich RP, Ehrlich AH (1972) Population resources environment. Issues in human ecology. W. H Freeman and Company, San Francisco, p 509

Eesley C, Lenox MJ (2006) Firm responses to secondary stakeholder action. Strateg Manag J 27:765–781

Etzioni A (1964) Modern organizations. Prentice-Hall, Inc, Englewood Cliffs, p 120

Falasca-Zamponi S (2011) Waste and consumption: capitalism, the environment, and the life of things. Routledge, New York, p 65

Freeman RE (1984) Strategic management. A stakeholder approach. Pitman Publishing Inc, Marshfield/London, p 276

Guerrero LA, Maas G, Hogland W (2013) Solid waste management for cities in developing countries. Journal of Waste Management 33:220–232

Haas W, Krausmann F, Wiedenhofer D, Heinz M (2015) How circular is the global economy?. An assessment of material flows, waste production, and recycling in the European Union and the world in 2005. J Ind Ecol 00(0):1–13

Heidrich O, Harvey J, Tollin N (2009) Stakeholder analysis for industrial waste management systems. Journal of waste management 29:965–973

Henry RK, Yongsheng Z, Jun D (2006) Municipal solid waste management challenges in developing countries-Kenyan case study. Jurnal Waste Management 26:92–100

Herayani FA (2011) Competency based human resources development analysis Bidang Kebersihan Dinas kebersihan dan Pertamanan Kota Cimahi Thesis Sekolah Tinggi Ilmu Administrasi Negara-Lembaga Administrasi Negara Bandung, p 228

Hoppe T, Sanders MPT (2014) Agricultural green gas demonstration projects. In the netherlands. A stakeholder analysis. Environ Eng Manag J 13(12):3083–3096

Joseph K (2006) Stakeholder participation for sustainable waste management. Habitat Int 30:863–871

Kastaman R dan Kramadibrata AM (2007) Integrated solid waste silarsatu reactor management system. LPM Universitas Padjadjaran, Bandung, p 167

Magness V (2007) Who are the stakeholders now? An empirical examination of the Mitchell, Agle, and wood theory of stakeholder salience. J Bus Ethics 83:177–192

Marshall RE, Farahbakhsh K (2013) Systems approaches to integrated solid waste management in developing countries. Journal of Waste Management 33:988–1003

Ministry of Environment (MoE) (2008) Domestic solid waste statistic of Indonesia 2008. Jakarta, p 23

Ministry of Environment (MoE) (2013) State of the environment report Indonesia 2012. Pillars of the environment of Indonesia, Jakarta, p 292

Mitchell RK, Agle BR, Wood DT (1997) Toward a theory of stakeholder identification and salience: defining the principle of who and what really counts. Acad Manag Rev 22(4):853–886

Muthmainnah L (2007) Menggugah partisipasi dan membangun sinergi: upaya bergerak dari stagnasi ekologis pengelolaan sampah (evocative participation and build synergy: an attempt to move from ecological stagnation waste management). Ilmu Sosial dan Ilmu Politik J 11(2):153–286

Neville BA, Bell SJ, Whitwell GJ (2011) Stakeholder salience revisited: refining, redefining, and refueling an underdeveloped conceptual tool. J Bus Ethics 102:357–378

Okumu JO, Nyenje R (2011) Municipal solid waste management under decentralization in Uganda. Jurnal Habitat International 35:537–543

Parent MM, Deephouse DL (2007) A case study of stakeholder identification and prioritization by managers. J Bus Ethics 75:1–23. https://doi.org/10.1007/s10551-007-9533-y

Reed MS, Graves A, Dandy N, Posthumus H, Hubacek K, Morris J, Prell C, Quinn CH, Stringer LC (2009) Who's in and why? A typology of stakeholder analysis methods for natural resource management. J Environ Manag 90:1933–1949

Scheepens AE, Vogtländer JG, Brezet JC (2016) Two Life Cycle Assessment (LCA) based methods to analyse and design complex (regional) circular economy systems. Case: making water tourism more sustainable. J Clean Prod 114:257–268

Seadon JK (2006) Integrated waste management-looking beyond the solid waste horizon. Waste Management Journal 26(12):1327–1336

Sejati K (2009) Integrated solid waste management using node, sub point, centre point system. Kanisius, Yogyakarta, p 88

SLHD (2008) Status lingkungan hidup daerah Kota Cimahi tahun 2008 (Regional environmental status of Cimahi City 2008), Cimahi

SLHD (2011) Status lingkungan hidup daerah Kota Cimahi Tahun 2011 (Regional environmental status of Cimahi City 2011), Cimahi

Srivastava PK, Kulshreshtha K, Mohanty CS, Pushpangadan P, Singh A (2005) Stakeholder-based SWOT analysis for successful municipal solid waste management in Lucknow, India. Waste Manag J 25:531–537

Sulistyorini L (2005) Pengelolaan sampah dengan cara menjadikannya kompos (Waste management by making compost). Kesehatan Lingkungan Journal 2(1):77–84

Suttibak, Samonporn dan Vilas Nitivattananom (2008) Assessment of factors influencing the performance of solid waste recycling programs. J Resour Conserv Recycl 53:45–56

Utami BD, Indrasti NS, Dharmawan AH (2008) Household solid waste management based on community: example from two communities in Sleman and Jakarta Selatan. J Trandisiplin Sosiologi, Komunikasi, dan Ekologi Manusia:40–68

Zurbrügg C, Gfrerer M, Ashadi H, Brenner W, Küper d D (2012) Determinants of sustainability in solid waste management – the gianyar waste recovery project in Indonesia. Waste Manag 32:2126–2133

Chapter 6
The Urban Solid Wastes Management in Cuautlancingo, Puebla, Towards a Circular Economy: Social and Economic Impacts of CE to the Region – Innovative Business Models

Luz del Carmen Díaz-Peña and Miguel Angel Tinoco-Castrejón

Abstract The purpose of this paper is to describe how Cuautlancingo, a local government of Puebla, México, manages circular economy (CE) activities related to urban solid wastes (USW). The final goal of those activities is to reduce potential sources of soil/water contamination and to preserve the environmental conditions. Due to the CE systematic approach, several stakeholders are required to integrate their efforts. Government and companies represent key stakeholders who can adopt the circular economy as a systematic thinking approach to make decisions accordantly. Hence, this paper was focused on trying to identify and describe the CE systematic approach in Cuautlancingo, as a case study. Three main points of attention were used for this analysis: (1) the wastes collection methods applied; (2) the type and conditions of waste management and (3) how does a circular economy perspective can influence other business or companies to recycle waste and generate more employment possibilities. The research methodology used was of qualitative nature. The main research method consisted of a questionnaire which was applied to staff members of the local government in Cuautlancingo. Aspects associated to the USW, such as generation, use and their treatment were the focus of the questionnaire. Some of the findings shed light on the poor CE implementation associated with USW management. However, some opportunities could also be identified, i.e. if the local government assumed the responsibility to organise and sell the recyclables, this would represent an additional income up to 366 USD per month. Consequently, it is perceived the need of redesigning the USW processes to retain the material value within the CE perspective.

Keywords Urban solid waste management · Cuautlancingo · Mexico · Circular economy

---

L. del Carmen Díaz-Peña (✉) · M. A. Tinoco-Castrejón
Instituto Tecnológico y de Estudios Superiores de Monterrey, Puebla, Mexico
e-mail: luzdiazp@itesm.mx; mtinoco@itesm.mx

M.-L. Franco-García et al. (eds.), *Towards Zero Waste*, Greening of Industry Networks Studies 6, https://doi.org/10.1007/978-3-319-92931-6_6

6.1 Introduction

Many scholars have written about the weak waste management system to cope with the global waste generation which yearly sums about 10,000 millions of tons. Less than half of them is collected and in the best circumstances also recycled. In the specific situation of Mexico, the per capita generation reaches 360 kg of waste per year (OECD 2015). Therefore, solid waste management can be linked to concepts like the CE in order to identify the opportunities to transform "waste" into "material" streams again.

In 2008, the Organisation for Economic Co-operation and Development (OECD) announced that there were 10,000 million tons of waste per year around the world and not even the half of them were conducted into a recycling treatment. Municipal waste represents only 10% of total waste, but its management and treatment can represent more than one-third of public sector financial efforts to abate and control pollution. Inappropriate waste management impacts human health and the environment due to soil and water contamination, air quality, climate, land use and landscape. In fact, the global biodiversity loss is projected to continue, with a further 10% loss by 2050 (OCDE 2015).

In Mexico, every year, the population generates 31.5 millions of tons of waste at their homes, stores and public areas, approximately 0.65% of it ends up in the streets as litter and 14.5% is burned up. According to the OCDE (2015), Mexico is the 25th country in the world that generates more kilogrammes of waste per capita: 360 kg of waste per year. In this setting, an approach that can contribute to the abatement of pollution and to the world climate change is the CE. The concept of the circular economy has gained increasing prominence in academic, practitioner and policy circles. Circular economies are achieved mostly through global recycling networks which are the primary means by which wastes are recovered as resources (Gregson et al. 2015).

Cuautlancingo was selected as a case in point to analyse the CE tenets because it represents one of the most important cities of the state of Puebla in Mexico, with similar demographic characteristics than other cities in the country. Even further, this case clearly displays recycling treatment practices of the urban solid wastes and its social and economic implications.

The collection, waste disposal, hygiene and cleanliness of marketplaces and other services are today's challenges of Cuautlancingo; if those are not attended in an efficient way, they will have a negative direct impact on the community quality.

The first part of the paper is an overview of the CE concept and principles, some of its benefits and the descriptive information of Cuautlancingo. In the second section, the research methodology is deployed including the data analysis methods. Results are presented and discussed in the third section to finally leave space to the conclusions and recommendations for further research about practical implications that may improve the management decisions that take circular economy principles into account.

The article starts by presenting an overview of the CE concept and some of its benefits, such as the improvement of economic results at the same time as reducing the use of resources by innovative ways. Increasing efficiency of resources management can indeed prevent environmental pollution and depletion and all in all contribute to climate change mitigation.

6.2 The Concept of Circular Economy (CE)

The idea of circular material flows, as a model for the economy, was presented in 1966 by Kenneth E. Boulding in his paper "The Economics of the Coming Spaceship Earth". Promoting a circular economy was identified as a national policy in China's eleventh five-year plan starting in 2006. The Ellen MacArthur Foundation, an independent charity, established in 2010, has more recently outlined the economic opportunity of a circular economy. As part of its educational mission, the foundation has worked to bring together complementary schools of thoughts and create a coherent framework, thus giving the concept a wider exposure and appeal.

Shi et al. (2006) defines the concept of CE as the interlinked manufacturing and service businesses. The promoters of CE seek the simultaneous enhancement of economy and environmental performance through collaboration by managing environmental and resource issues. By working together, the community of businesses seeks a collective benefit that is larger than the sum of the individual benefits of each enterprise, industry and/or community.

CE promotes a harmoniously economic development pattern with the earth. The main purpose of CE is to organise the economic activities to a close-loop process of "resource-production-consumption-regenerated resource". All materials and energy can be used rationally, efficiently and continuously in sustainable economic cycles; hence, the harmful effect to the natural environment can be reduced to a minimum. The principles of CE are based but not exclusively related to the "reduce, reuse and recycle" (3Rs) (Shi et al. 2006).

The implementation of CE strategies is expected to achieve an efficient economy rate while discharging fewer pollutants, although the CE strategy requires complete reform of the whole system of human activities, which includes both production processes and consumption activities. It has been widely explained that CE could help to improve resource productivity and eco-efficiency, reform the management of the natural resources and achieve sustainable development standards (Yuan et al. 2006). Activities over the past several years, however, clearly show that CE is emerging as an economic strategy rather than a purely environmental preservation approach. The major objective of the government is to promote the economic and environmental dimensions of the sustainable development concept (Yuan et al. 2006).

Another CE definition was given by "The Ellen MacArthur Foundation", which is: "CE is one that is restorative by design, aiming to keep products, components and materials at their highest utility and value at all times, distinguishing between technical and biological cycles. The distinction between biological and non-biological products has the purpose to maximise the materials value recovery. It is based on renewable resources – either grown or circulated within the economy at high value". The circular economy has no wastes: only resources designed to be feedstocks for other products (The Ellen MacArthur Foundation 2013).

The development of the CE has brought with it the conclusion that this approach would be more efficient for the resource productivity if it focuses on adjusting industrial structure, developing new technology and reforming industrial policy, rather than just recycling waste.

In the same line but at the consultancy level, "The Green Gain" firm is a specialist in the circular economy, innovation and enterprise. It has experience of working with organisations on sustainable waste, water and energy efficiency and helps them to respond to changing markets and increasing resource prices through a range of strategic and practical approaches. Moreover, Green Gain (2016) presented some key features of circular economy thinking, which are:

- **Whole system design:** an effective circular economy for a product is much more likely to be achieved if everyone involved are incentivised towards the same outcome.
- **Design for end-of-life recovery**: the repair, refurbishment and remanufacture become more commercially viable.
- **New business models:** selling the function of a product rather than the product itself means that the supplier retains ownership of the product and is incentivised to design for durability and end-of-life recovery.
- **Diversity:** a wide variety of circular economy solutions applied at different scales will provide a resilient and adaptable foundation.
- **Biomimicry:** copying systems used in nature – seeing ecosystems as a template for a circular economy and its elements.
- **Innovation:** the process includes invention, creativity, experimentation, entrepreneurship and collaboration.

6.2.1 Global Perspective of CE

In 2015, a report was released entitled Towards the Circular Economy: Economic and Business Rationale for an Accelerated Transition. The report, commissioned by the Ellen MacArthur Foundation and developed by McKinsey & Company, was the first of its kind to consider the economic and business opportunities for the transition to a restorative, circular model. "The Ellen MacArthur Foundation, SUN, and McKinsey have identified that by adopting circular economy principles, Europe can take advantage of the impending technology revolution to create a net benefit of €1.8 trillion by 2030, or €0.9 trillion more than in the current linear development path" (Ellen MacArthur Foundation 2013).

In March 2014, "Resource" did the first large-scale event in London for the CE with over 11,000 attendees from across the globe representing all the major stakeholders. The launch of such an event signals the rise of the topic, and it will act as an enabler for business to transit towards more circular business models. "Resource" is built around the needs of companies that want to save and make money through reducing waste, recycling, recovery and repair, but now these companies are looking to take the next step towards resource efficiency and the circular economy. Resource connected professionals from across the textiles, automotive, food, retail, chemicals, construction and electrical industries to help them to learn about the circular economy, network and find new partners and solutions. The 2016 Conference brought together leaders from across Europe to share the latest thinking, insight and analysis around corporate resource efficiency and circular economy modelling (Resource 2015).

6.2.2 Mexican Perspective of CE

Regarding the circular economy in Mexico, there is no much information or statistics about it (2015). The perception is that it is a relatively new concept, and the large companies are mainly the ones that are adopting important environmental standards as part of their voluntary policies. Some improvement of their CE's processes through the ISO 14000 certification has been reported, for instance, Heineken Mexico announced its admission at the "CE 100 Programme" created by the Ellen McArthur Foundation. On the other hand, very few of the micro and small companies decided to add environmental care as part of their regular activities (Carrillo et al. 2010).

From the legal-institutional viewpoint, the environmental policies in Mexico are framed in the "Political Constitution of the United Mexican States" and the "Ministry of Environment and Natural Resources" (SEMARNAT[1]). This latter is on charge to develop and implement more specific laws, norms and create institutes. The most important laws related to CE are Organic Law of the Federal Public Administration, the General Law of Ecological Balance and Environmental Protection and the General Law for the Prevention and Integral Management of Wastes. These laws state that each federative and local entity in Mexico has the responsibility to create, manage and evaluate its own environmental legislation in order to preserve and restore the ecological balance of the physical environment. In the last decades, it has been seen that these entities do not have neither capacities in place to face the environmental challenges nor the budget and infrastructure (Carrillo et al. 2010). More recently, a real effort for having an institutional structure and legislation is perceived. The main environmental issues they have been working at are potable water, drainage, sewage system, collection and final waste destinations and the creation of ecological reserves zones (Carrillo et al. 2010).

[1] Acronym in English

From the practitioner's perspective, through this research, it could be identified three organisations in Méxcico that have used the circular economy: "Zero Waste", "Urban Island" and "Bio bag System". A short description of each of them is mentioned as follows.

"Zero Waste" is a company that collects industrial and landfill's wastes. It also recycles and transforms them into new materials like sheets, carpets, daises or scholar's tools. These products are again integrated in the consumption chain as a fully recycled products. In the case of "Urban Island", they collect rainwater through a piping system and bring it to homes, schools and isolated zones. The interesting point here is that the main resource is from a natural source. This organisation is located in Mexico City and has more than 2000 catchment systems that benefit more than 10,000 persons at country level. In the case of "Bio bag System", this company produces, installs and sells biodigesters that treat organic wastes such as manure. With a decomposition process, the wastes are used for biofuels and compost production. These biodigesters are installed mainly in farms, schools and rural areas for more than 12,000 users.

6.3 Urban Solid Wastes (USW)

As definition, wastes are those materials and products whose owner throw away and that stay in a solid or semi-solid, liquid or gas state in a container or tank. They can be valued or required to a specific final treatment. According to their characteristics and origins in Mexico, they can be classified into three groups: (1) hazardous wastes, (2) special use wastes and (3) urban solid wastes (organic and inorganic) (General Law for the Prevention and Integral Management of Wastes 2003).

The world population is growing at a rate of 1.6% per year (country meters 2015); thus, the urbanisation and urban wastes are also growing and contaminating the air and the groundwater when they are not properly managed. At the same time, the natural resources are becoming depleted. In this sense, all the individuals need to be conscious about these social and environmental problems. In a minor scale, counties and town councils (boroughs) have a direct, immediate and unavoidable responsibility towards the environment. A few years ago, they were not regarded as priority functions of the counties. At present, they have not only a basic legal obligation but the social commitment of the local government to the population (Pichardo 2009).

The urban solid wastes (USR) are generated in houses and shelters, streets and public places as a result of the material consumption that were used (General Law for the Prevention and Integral Management of Wastes 2003). These kinds of wastes have a very slow decomposition because of their chemical characteristics, and even though most of them are from a natural origin, they are not biodegradable. Generally, the USW are recycled through artificial and mechanical methods as it is the case of cans, glasses, plastics and gums (Mantra n.d). In 1998, 85% of the USR in Mexico were collected and in 2011 93%, but only 4.8% of them were recycled (SEMARNAT 2012a, b).

Table 6.1 USW in Mexico 2014 (SEMARNAT 2014 & compiled by authors in 2016)

State	Tons per day
Mexico City	16,486.55
Mexico State	13,014.29
Jalisco	6.944.16
Veracruz	6030.15
Puebla	4330.25
Michoacán	4187.8
Nuevo León	4042.19

Table 6.2 Wastes type in Mexico 2012 (SEMARNAT 2014 & compiled by authors, 2016)

Wastes type	Thousands of Tons, 2012	Percentage
Paper and cardboard	5822.82	13.83%
Textile	602.06	1.43%
Plastics	4584.99	10.89%
Glasses	2475.66	5.88%
Metals: aluminium	728.38	1.73%
Other metals	719.87	1.71%
Food, garden and other organic waste	22,070.27	52.42%
Disposables	5098.7	12.11%
Total	42,102.75	100%

The SEMARNAT informed in 2014 that Puebla occupied the 5th place of USW daily generation in Mexico; there are 31 states and 1 federal district in the country. In Table 6.1, the seven states that generate the most USW in Mexico are indicated.

From the total of 1894.35 miles of tons, 1609.45 are casted in controlled places (landfills) and 284.9 are casted in no controlled places and used for recycling processes. In average, every year, the urban solid wastes quantity in Puebla increases 2.91% for the last 10 years, in comparison with the 2.78% national average (SEMARNAT 2012a, b).

The above indicator is an important element of analysis, since it points out that there is great to do with the USR management in Puebla, specifically for the recycling part. Few statistics on the waste composition can be found, one of the latest (2012) is displayed in Table 6.2. This type of information is relevant in order to define what waste streams are potentially suitable for either any or the combination of the 3Rs (CE tenets). Table 6.2 shows that aside from the organic waste, more than 30% of the wastes in Mexico come from materials than can be recycled, e.g. paper, cardboard and glass.

Also relevant to say is that in 2011, paper and cardboard were the first recycled materials in the country (42%), followed by glass products 29%) and plastics (28%) (SEMARNAT 2014).

6.4 Cuautlancingo Case Description

Cuautlancingo Township is located in the west central part of the Puebla state, with 79,089 inhabitants (National Institute of Statistics and Geography 2010). The municipality is bordered on the north by the state of Tlaxcala, the south with the municipalities of San Pedro Cholula and Puebla, the east with the municipalities of Tlaxcala and Puebla and the West with the municipality of Coronango.

The state of Puebla has 16 federal districts, and Cuautlancingo is located in the number 10 with centralised management in San Pedro Cholula.

Cuautlancingo has an area of 3317 square kilometres. With a location in the Southern part of the Valley of Puebla, it has a flat topography and a few steep declines towards the Atoyac River, which is one of the most important watersheds in the state. It has three counties: La Trinidad Santorum, La Trinidad Chautenco and San Lorenzo Almecatla (INAFED 2015).

The selection of Cuautlancingo for this research was based on the presence of one of the most important industrial parks in Puebla (Finsa) with around 24 companies, including Volkswagen (Finsa 2015). Besides, the municipality has a solid waste management under the circular economy.

6.5 Methodology

The methodology used is of qualitative nature, and this is here further described in this section to explain how the information was obtained and analysed. The main stages to gather the information were (1) asking the municipality staff for interviews, (2) questionnaire design, (3) semi-structured interviews application, (4) analysis of information, (5) additional meeting to clarify questions with the people interviewed from the municipality and (6) a preliminary report ("*As Is*" *process mapping*[2]) elaborated, to be further developed and finalised later on.

As here above-mentioned, interviews with the governmental authorities responsible of the environment policy were carried out at different stages of this research and with different purposes. The first one had an exploratory character with the purpose to learn about the urban solid waste generation and its management. From this meeting, it was identified that the urban solid waste treatment is one of the main concerns of the municipality and that the local government has no solid waste management system at all. Because of this, the processes during the waste collection, separation and disposal are not well defined yet, and it is seen as an opportunity area. Hence, the main focus of the diagnosis was on the nonhazardous-inorganic solid waste treatment to delimit the scope of this research.

[2] The "*As Is*" diagram describes the present state of the organisation's process, culture and capabilities. The main purpose is to help to identify where the improvements areas are as potential starting points.

In order to perform the preliminary diagnosis, it was required to describe a set of questions to gather information from the Cuautlancingo municipality. In the second meeting, this questionnaire was personally applied not only to the head of the environmental department but also to the environmental council. The involvement of different waste operation areas provided an enriched variety of responses at that moment. The information was put together in order to be classified and analysed based on each process and on the way solid wastes could be recycled, reused or disposed. The information was also assigned for a further analysis focused on some other components to be considered in a circular economy (CE) loop.

By the time the first set of information was analysed, it was necessary to perform a check-up to exclude any potential misunderstanding and to clarify the issues that could be arisen during the information analysis stage. This action took place during one more meeting with the presence of multi-actors: the mayor of Cuautlancingo, the environmental council, the head of the environmental department and the waste operations department head. During this meeting, some more pieces of information were provided under request to round up the information that created a better understanding of prior inputs.

Based on the information provided, it was possible to identify the "*As Is process mapping*" to obtain a general idea on how many steps and interfaces are presented along the process of recollection, segregation and waste disposal.

6.6 CE Practices Application in Cuautlancingo Municipality

From the interviews and meetings with the local authorities of Cuautlancingo, in particular the one with the environmental councillor, several policy goals were mentioned regarding the implementation of concepts such as environmental and social sustainability. Circular economy was not per se named, but the CE tenets associated with waste management and economic and social value creation were clearly indicated. Even further, sustainability was regarded as the welfare improvement of the Cuautlancingo community (families and businesses). The plans and projects associated with the circular economy principles applied to the solid waste management founded are the collection of separated wastes; the recycled actions, awards and public recognition for companies that are socially responsible and care about the environment and territorial reserve; educative campaigns to increase environmental awareness; dissemination of best practices; etc. In a more descriptive manner, the plans and projects are described in the following sections.

6.6.1 Separated Waste Collection

The local government has six collection trucks, and they collect three times per week the municipal nonhazardous wastes, except for the Volkswagen plant and the industrial Park Finsa, because such industries hire private companies for the same purpose.

The homes and stores are responsible for the in situ first separation of organic and inorganic materials. As for the inorganic portion, several materials need to be separated according to their reintegration possibilities to the market; the materials are cardboard, glass, polyethylene terephthalate (PET) and cans (aluminium). Every Tuesday, Thursday and Saturday, the trucks collect those materials and drive them to the Panotla landfill, in the state of Tlaxcala.

In a monthly basis, 1.0 ton of cardboard, 1.5 tons of PET, 0.5 ton of glass and 0.3 ton of aluminium are collected. The costs related to the collection activities sum up to the amount of 2659.26 USD per truck, including the driver salary, gas and maintenance. This information is resumed in Table 6.3, and the monetary calculations were based on a currency exchange rate of 18.00 MXN/USD.

6.6.2 Actions of Recycling

From the collected materials, only 16% of each type is actually recycled; the rest are mixed with other wastes when they arrive to the landfill. The persons who live in the surroundings pick up the wastes, clean them and sell them to a third "actor" for their reuse. They can sell them per ton unit as follows: cardboard $60.60 USD[3]/ton, PET $260.60 USD/ton, glass $18.18 USD/ton and aluminium $909 USD/ton.

If the local government would take the responsibility to organise the recycling of the here above-mentioned materials, this would represent an additional income for the local administration. Table 6.4 shows some monthly estimations done at this regard if the municipality decided to sell directly those collected materials.

The geographical domain of the Cuautlancingo municipality includes an ecological park whose name is "El Ameyal". The park's surface corresponds to 7 hect-

Table 6.3 Collected materials and associated monthly costs (Authors contribution)

Product	Cardboard	PET	Glass	Can
Quantity collected tons/month	1	1.5	0.500	0.300
Payroll USD per 6 trucks/month	$1600			
Gas USD per 6 trucks/month	$8800			
Maintenance/month USD	$5555			

[3]All the currency conversions from Mexican (MXN) pesos to USD were done with the 18.00 MXN/USD exchange rate.

Table 6.4 Potential income according to sales percentage

Product	Carton	PET	Glass	Can	Total USD
Selling 16%	$9.69	$62.54	$1.45	$43.63	$117.31
Selling 50%	$30.30	$195.45	$4.545	$136.35	$366.64
Selling 100%	$60.60	$390.90	$9.09	$272.7	$733.29

Table 6.5 "The Ameyal Park" wastes management 2014–2015 (Adapted from Cuautlancingo municipality 2015)

2014		July 2015		
Wastes	Total (tons)	Wastes	Total (tons)	Total (tons)
Pet (1)	217.5	Pet (1)	348	565.5
Carton	121.5	Carton	78	199.5
Glass	136	Glass	130.9	266.9
Can	19.5	Can	22.5	42
Aluminium	23.5	Aluminium	15.1	38.6
Others	270.8	Others	52.8	323.6
Organic	65.8	Organic	50	115.8
Recycled	**885.7**	**Recycled**	**697.3**	**1583**
Waste	**1371**	**Waste**	**2001.20**	**3372.20**
Total	**2256.70**	**Total**	**2698.50**	**4955.20**

ares with some facilities such as 600 lineal metres track, exposition centre, children's playground, picnic area and theatre. The government of Puebla donated 160 trees to the park, and Volkswagen supported the planting of more than 1000 trees in the park's green areas. The Volkswagen's executives participated in those activities who also donated pumping and watering equipment for tree maintenances. Additionally, they provided trash cans for the park.

The urban solid wastes management in the park began in August 2014. They had from the beginning a strict weight control of each type of generated waste. Some of them are displayed in Table 6.5.

All the materials are stored in the recycling park storehouse for their sell at the end of the year to a recycling company.

Additionally, the local government has an agreement with the Autonomous University of Puebla (BUAP) that consists in collecting the organic wastes and leaf litter, from the Ameyal Park, to produce fertiliser.

6.6.3 *Awards and Public Recognition and Territorial Reserve*

The local government of Cuautlancingo awards and recognises the companies that are socially responsible and care about the environment; some of the awarded actions are directly related to the CE perspective. For instance, in April 2015, some schools were awarded because of their projects associated with materials recycling and environmental care. The award is called "Clean School" and is given to the 71 Technical School and The Gabriel Alatriste Junior High School, both located at San Lorenzo Almecatla municipality. These schools produced fertiliser with their own organic waste and with recycled cardboard handcrafted little home's souvenirs, palapas[4] and school supplies. As an incentive, the students received backpacks and football balls for their important contribution.

Furthermore, the City Inn Hotel carries out sustainable actions by using recycled flatware and the "Elektro Controles y Motores de Puebla" company recycled the wastes making objects for internal use.

Cuautlancingo invested 4.375 million dollars in a territorial reserve called Quetzalcoatl-Chautenco, 1.25 million in land and 3.125 million in infrastructure. This reserve includes terraces, the Ameyal Park, a wealth centre and other services that provide support to the community. The management of the reserve relies on its own inspection office that helps for organising and planning all the services.

6.6.4 Workshops to Increase Environmental Awareness

The municipality of Cuautlancingo considers very important to raise awareness and educate the population about circular economy principles and its implementation and benefits. Therefore from 2015, the authorities delivered free workshops in their facilities on how to separate wastes, to recognise responsible products, to produce fertilisers and to construct vegetable gardens. The number of attendees in 2015 was 497 people, expecting to receive more each year.

6.6.5 Upcoming Projects

Because the municipality of Cuautlancingo doesn't have a centre to transfer the wastes where the USW can be casted after collection, the construction of one is in the municipality planning. The local authorities have met several times and came up with the plan to invest 25,000 USD in the transfer centre facilities. The objective is to create an inter-municipality landfill with the intervention of nine municipalities and to run it with renewable energy sources. With such project in mind, the savings could reach 30% of the cost for each municipality involved. The purpose is to have a space where the transporters can carry on the wastes and separate them for their respective uses.

[4] Palapa is an open-sided dwelling with a thatched roof made of dried palm leaves. It is very useful in hot weather and in the coast of Mexico.

6.6.6 *Best Practice: Volkswagen*

As a best practice to be followed by the local government or other companies, Volkswagen company was visited in order to learn about its activities related to circular economy and sustainable processes for the USR management. The company, which is certified as "socially responsible", operates based on an environmental performance that includes materials recycling, water treatment and renewable energies. The company "Think Blue" project targets all those aspects together to contribute to the environmental improvement, also by considering the people awareness on their well-being standards. The Think Blue strategy is shown in Table 6.6.

In 2014, the environmental benefits attributable to the "Think Blue Project" can be here summarised in three main aspects: (1) the use of energy decreased 19% in comparison with 2010, (2) the water consumption reduced 9% and (3) the waste generation was reduced 96% compared to 2010. In addition to that, the carbon dioxide decreased 16% in the same period. (Volkswagen plant 2015).

Table 6.6 Think Blue strategy (Adapted from Volkswagen 2015)

Products and solutions	Ecologic awareness	Cooperation and initiatives
Blue motion technologies	Education	Think Blue nature
TSI & DSG	Visits to universities	"Itza-popo" project
Turbo diesel	Environmental education	"Sierra de Lobos"
Hybrids		"For love to the planet"
Electrics		
Think Blue finance		
Think Blue factory	Training	New projects
Sustainable production	Training and technical advice to employees	Tire recycling
		New water treatment plant
		Technical-social projects with sustainable technology
		Eolic project park
	Diffusion/communication	
	Newspaper, magazines	
	Switch information with other assemblers	
	Diffusion with local and national media	
	Reports about sustainability	
	Environmental ideas	
	Tips for saving energy	

6.7 Conclusions

Based on the evidence gathered during this research, it can be said that the government of Cuautlancingo cares about its environment and performs actions to contribute to the welfare and quality of life of the community. The services delivered by the municipality have a direct effect on the local environment. Drinking water, domestic drainage, waste disposal, wastewater treatment, hygiene and cleanliness of marketplaces and other services are today's challenges of Cuautlancingo. In addition to that, the urban solid wastes management in Cuautlancingo is lagging behind when comparing it with what circular economy promotes. Consequently, it is convenient that the municipality redesigns processes and offers products and services more aligned to the circular economy tenets that can bring additional income.

Even further, investing in greening can generate employment and sustainable supply chains for the products. The more companies recycle and reuse the USW, the higher productivity level of the municipality can be achieved. Furthermore, based on the leader companies' best practices, other local governments and companies may benchmark and adjust such practices to implement actions that lead to a waste management efficiently.

The authors suggest that Cuautlancingo continues registering and analysing indicators in order to assess circular economy in a more accurate way. Some indicators that can measure the progress of the greening of sectors are the resource consumption rate (per capita), the waste generation rates and how much of it is segregated and recycled and the proportion of waste used for energy recovery.

To further promote the circular economy, the local government shows willingness to work hard to formulate regulations and policies in order to ensure the environmental care and responsible acting of the citizens. It is very important that Cuautlancingo continues working with the environmental education campaigns and with the awareness of the wastes management opportunities for the community.

As a reflection on the way this research was carried out, the authors noticed that the government officers are interested in this kind of grouping work where stakeholders meet to discuss all the needed key aspects in benefit of the city. They often showed an enthusiastic and cooperative attitude in order to answer the questionnaire applied and were conscious that there is still so much to improve in the solid residues management, especially in the recollection phase.

The future research work of this paper could compare the USW management of some other municipalities with similar characteristics as Cuautlancingo, for instance, with an industrial park or the same population, all that with the intention to analyse the possible generalisation of the findings of this study in point. The contextual conditions would need also to be further described in case comparison among different political entities takes place. This implies the review of this case study if there were some respective new initiatives of laws in Puebla (state).

6.7.1 Social, Economic and Environmental Implications

The present social conditions of the municipality are becoming more favourable in benefit of its population, with a clean and organised city and a better quality of life, and the relations between people, small business and companies can improve and contribute in the sustainability of the specific location. The most important element in one economy or cluster are the people, and with acceptable social conditions, they can act in a better way.

It is important that the local government continues implementing incentives for those companies that do something for the environmental care. Local governments can reform their laws in order to reduce either the taxes or the operating permissions costs for companies. Their focus should be on some alarming environmental threats are ozone layer depletion, atmospheric pollution, natural species extinction, acid rain and desertification (Navarro, 2013).

Cost recovery from improved waste management and avoided environmental and health costs can help to reduce the financial pressure on governments. Private sector participation can also significantly reduce the costs as well as enhance service delivery. Micro-financing, other innovative financing mechanism and international development assistance may in addition be tapped to support operational costs for waste treatment. Finally, a range of economic instruments can serve as incentives to green the sectors, and their implementation could be combined with regulations to set a minimum of safety standards to protect labour conditions.

References

Carrillo G, Constantino T, Roberto, Roldán, Alejandra H (2010) Incentivos de la política ambiental para ecología industrial en México. Innovación y ecología industrial. http://docplayer.es/10729896-Incentivos-de-la-politica-ambiental-para-ecologia-industrial-en-mexico.html. Accessed 12 Aug 2015

Countrymeters.info (2015) Población mundial. http://countrymeters.info/es/World. Accessed 23 Sept 2015

Ellen MacArthur Foundation (2013) Towards the Circular Economy: an economic and business rationale for an accelerated transition. http://www.ellenmacarthurfoundation.org/business/reports/ce2012. Accessed 19 Aug 2016

Federalism and Municipality Development National Institute (Instituto Nacional para el Federalismo y Desarrollo Municipal INAFED) (2015) Encyclopedia of the municipalities of Mexico. Cuautlancingo. http://www.inafed.gob.mx/work/enciclopedia/EMM21puebla/municipios/21041a.html. Accessed 2 May 2015

Finsa (2015) FINSA copyright. http://www.finsa.net/?parques=finsa-puebla. Accessed 2 May 2015

Green Gain (2016) Circular economy. http://www.greengain.co.uk/circular-economy.html. Accessed 10 Sept 2016

Gregson N, Crang M, Fuller S, Holmes H (2015) Interrogating the circular economy: the moral economy of resource recovery in the EU. Econ Soc 44(2):218–243. http://www.tandfonline.com/doi/full/10.1080/03085147.2015.1013353 Accessed 7 July 2015

Mantra (n.d.) RESIDUOS ORGÁNICOS E INORGÁNICOS. http://www.mantra.com.ar/contecologia/organicoseinorganicos.html. Accessed 8 Sept 2015

Navarro GF (2013) Responsabilidad Social Corporativa: Teoría y práctica, 2nd edn. ESIC, Madrid, pp 163–210

Organization for Economic Co-operation and Development (OCDE) (2015) Waste. https://data.oecd.org/waste/municipal-waste.htm. Accessed 11 Aug 2015

Pichardo PI (2009) Responsabilidades municipales en materia municipal. *Convergencia.* ISSN 1405-1435 vol 16 No. 49 Toluca, México

Resource (2015) http://www.resource-event.com/welcome. Accessed 12 Aug 2015

SEMARNAT (2012a) Base de Datos Estadísticos-Badesniarn. Consulta Temática 2012. http://dgeiawf.semarnat.gob.mx:8080/approot/dgeia_mce/html/mce_index.html?De=BADESNIARN. Accessed 3 Aug 2015

SEMARNAT (2012b) Residuos. http://app1.semarnat.gob.mx/dgeia/informe_12/pdf/Cap7_residuos.pdf. Accessed 3 Aug 2016

Shi L, Xing L, Bi J, Zhang B (2006) Circular economy: a new development strategy for sustainable development in China. Available via http://www.webmeets.com/files/papers/ERE/WC3/161/Circular%20Economy%20in%20China.pdf. Accessed 28 June 2015

The General Law for the Prevention and Integral Management of Wastes (2003) Fifth article. Mexico

Volkswagen (2015) http://www.vw.com.mx/es/mundo-volkswagen.html. Accessed 2 May 2015

Yuan Z, Bi J, Moriguichi Y (2006) The circular economy: a new development strategy in China. J Ind Ecol 10:4–8. https://doi.org/10.1162/108819806775545321

Chapter 7
A Massive Urban Symbiosis: A Preliminary Review of the Urban Mining Pilot Base Programme in China

Yanyan Xue, Hans Bressers, and Zongguo Wen

Abstract Waste recycling helps to establish a circular loop of resource flow between production and consumption, achieving a certain symbiosis between the industrial and urban sector. Since more and more resources are accumulated in the urban sector, urban mining as form of waste recycling in a massive way becomes an outstanding way to achieve industrial and urban symbiosis. In 2010 China initiated a national urban mining pilot base (UMPB) programme with the objective of developing the recycling industry and relieving environmental and resource constrains. This study aims to provide policy review of the programme. We find that the UMPB programme was developed from past circular economy policies and attains legacy assurance from current laws and national plans. But this did not formulate a perfect governance context for its implementation. A multi-ministerial cross-management network led to policy conflicts, and recycling-oriented legislation remained absent. These became the main barriers for the good implementation of those urban mining pilots. Comparing with the eco-town programme in Japan, it shows that both programmes share some similarities of partial policy objectives but also show variety in the scope of urban symbiosis due to the different problems they focus on and the slightly different policy objectives under the different economic and social development phases.

Keywords Urban mining · Urban symbiosis · China · Recycling industry

Y. Xue (✉)
School of Environment, Tsinghua University, Beijing, China

H. Bressers
University of Twente, Enschede, The Netherlands
e-mail: j.t.a.bressers@utwente.nl

Z. Wen
Environmental Planning and Management, Tsinghua University, Beijing, People's Republic of China

Research Center of Circular Economy, School of Environment, Tsinghua University, Beijing, People's Republic of China
e-mail: wenzg@tsinghua.edu.cn

M.-L. Franco-García et al. (eds.), *Towards Zero Waste*, Greening of Industry Networks Studies 6, https://doi.org/10.1007/978-3-319-92931-6_7

7.1 Introduction

Resource supply is an international issue for many countries in the process of urbanisation and industrialisation. UNEP predicted that the global demand of in-use metal stocks would be increased by 3–9 times if the world total population were to enjoy the same levels of use as industrialised countries (UNEP 2010). China's outstanding economic growth pattern is resource intensive, having become "the world's factory". Ministry of Land and Resources (MLR) report shows that China's external dependence on imported iron, copper and aluminium is up to 51.2%, 72% and 47.9%, respectively (MLR 2016).

Recycling is regarded as one important solution to ease the resource depletion in many counties. Recycling rates of 18 metals are estimated exceeding 50% and those of glass, plastic and other packaging materials even higher than 90% (Graedel 2011). Moreover, recycling can also save energy and avoid environmental pollution compared with virgin resource exploitation. Such extraction of secondary resources from urban metabolism is named as urban mining (Krook and Baas 2013; Krook et al. 2010). Since urban theorist Jacobs noticed that the cities will become a huge, rich and diverse mine of raw materials (Jacobs 1969), urban mining has become a metaphorical term for resource recovery from the technosphere. Literature suggests that urban mining has various denotations in different contexts. A broader concept includes the landfill mining and mining the tailings, the slags, the dissipations, the hibernations and in-use stocks (Johansson et al. 2013). The urban mining potential of copper and iron will attain 8.1 and 711.6 million tons, respectively, in 2040. The substitution rate (secondary metals substituting primary metals) can increase by 25.4 and 59.9% compared to the status in 2010 (Wen et al. 2015).

In 2010, the Chinese government initiated a national programme to establish 50 national urban mining pilots base (UMPB) in China. In its official notification, urban mining is defined as recycling waste materials from the major seven waste stream, including the electronic equipment, cables, communication facilities, vehicles, household appliances, electronic products, packaging and scraps (NDRC 2010). These are obviously classified as end-of-life products of in-use stock. The so-named bases are often referring to industrial parks hosting large-scale waste-recycling plants including pretreatment, processing and products manufacturing. In this process, massive resources are recovered from waste and transited into secondary materials and even new products. In 2012, 29 bases recycled 24 million tons of waste and produced 16 million tons of secondary materials with a total market value of 247 billion RMB.[1]

Such massive waste recycling establishes a circular loop of material flow between production and consumption and between industry and urban to enable symbioses emerging in the urban sector. The urban symbiosis concept is first introduced by Van Berkel, taking the Japanese eco-town case as example (Van Berkel et al. 2009), and was followed by several more publications on the same subject (Dong et al. 2014;

[1] Equivalent to 35.9 billion USD at an exchange rate of 1 USD to 6.88 RMB on 23 May 2017

Geng et al. 2010). Contrary to the intensively studied industrial symbiosis (Chertow 2000, 2007; Golev et al. 2014; Shi et al. 2010; Van Berkel 2004), the study of urban symbiosis is still at an early stage. This paper aims to bring the Chinese urban mining base case into this rarely studied field and answer the following questions: *What are the typical features as a massive urban symbiosis policy that the Chinese urban mining base programme represents, and to what extend can it be improved?*

Data was collected and analysed by reviewing documents from government and by interviewing policy makers and the pilot base managers. The paper is organised as follows: after this introduction, Sect. 7.2 profiles recycling industry in China; Sect. 7.3 introduces the urban mining programmes including the management frameworks and the progress of its implementation. Section 7.4 presents a policy analysis including the policy evolution and governance context and analysing those with the help of the governance assessment tool (Hans Bressers et al. 2016). Section 7.5 compares the UMPB programme with the Japanese eco-town programme, and conclusions are in Sect. 7.6.

7.2 Development of China's Recycling Industry

Since the 1980s, some farmers began to engage in recyclable waste resources collection and recycling by establishing recycling workshops in their backyards. They traded and collaborated with each other and gradually cultivated a leading industry in the county, contributing to the local economy and employment. For example, the plastic waste-recycling industry in the Gengche town in the Jiangsu province accounted for 80% of the GDP in total (Zhao 2011). It also created jobs for the surplus rural labour force of the towns; 70% of the residents of Gengche town worked on waste collection and recycling business, and people from adjacent towns also joined in the industry; altogether 60,000 jobs were created (Zhang 2010). Table 7.1 lists some typical recycling industry township/county in China.

The spontaneously developed recycling industry in towns bears some problems. Firstly, there is no layout plan for the household workshops; the wastes are randomly

Table 7.1 Some typical township or county recycling industry aggregations (Authors' contribution)

Town/county	Recycling industry	Recycling amount per year (10^4 tons)	Workers involved (10^4)
Wen'an, Hebei	Plastic	200	10
Jieshou, Anhui	Plastic, lead acid battery	200	5
Dazhou, Henan	Metals	380	4.5
Guiyu, Guangdong	Imported WEEEs	220	6
Ziya, Tianjin	Imported hardware	150	2
Gengche, Jiangsu	Plastic	300	3

piled up in open air and dismantled and processed with poor equipment, imposing environmental and health risk to the workers and local residents. Wen'an in Hebei province, a county 150 km from Beijing, became the largest waste plastic assembling site in the past 20 years but also turned from a bucolic agricultural region to a bustling, crowded, dirty, stinky and noisy environment (Minter 2013). Manually dismantling imported WEEEs in Guiyu of Guangdong province caused severe environmental problems and even poor health and high neonatal mortality (Song and Li 2014; Xing et al. 2009; Xu et al. 2012).

Secondly, recycling in China is mainly driven by market forces and economic profits. Waste metal and paper/cardboard have higher market value leading to high collecting and recycling rates, while waste plastic and glass have lower recycling rates, and the collecting and recycling of waste compact fluorescent lamps remain deficient due to little profits and the absence of regulations. Thirdly, because most recycling plants cannot afford advanced technology and equipment, they often only reclaim high-value metals and discard the compound metals that are difficult to be separated, resulting in the overall recycling rate staying very low. The WEEE's reclaim rate is only 30% in China, while that of European countries can be as high as 75%. These are not helpful for resource saving and circular economy development set by the central government; thus low-end recycling industry needs improvement.

7.3 Urban Mining Pilot Base (UMPB) Programme in China

In 2010, the China National Development and Reform Commission (NDRC) and the Ministry of Finance jointly initiated the UMPB programme. The objective stated in the official document is "to implement the Circular Economy Promotion Law, to promote the recycling industry development and help to relief the resource and environmental bottleneck constrains in China". The goal is to support 30 (later upgraded to 50) national urban mining pilot bases. Through this process, it intends to promote the key waste streams recycling at a large scale with high-value production, to develop and spread advanced recycling technology and to explore the urban mining model and policy mechanism. For this purpose, the programme prescribes seven requirements for an ideal urban mining base: systematic waste collection network, proper industrial chain, upscaled recycling of materials, advanced equipment, shared infrastructure, collective environmental facilities and a standard management and operation system (Xue et al. 2017).

The prediction of the urban mines' potential and waste stream tracing as well as the spatial distribution of the recycling plants needs in-depth research, but some studies indicate that it is very promising. The urban mining potential of copper and iron will attain 8.1 and 711.6 million tons, respectively, in 2040. The substitution rate (secondary metals substituting primary metals) can increase by 25.4% and 59.9% compared to the status in 2010.

7.3.1 Selection of the Pilot Bases

The pilot bases are selected from the recycling industry parks, and only those parks with annual recycling capacity above 0.3 million tons are eligible to make application. The selection process consists of four steps: (1) provincial DRCs recommend local applicants; (2) the applicants draw up and submit their national urban mining pilot base action plans according to the guideline issued by NDRC. The guideline sets a unified format for the action plans and emphasises the park's status quo analysis, the target of recycling capacity and the newly added waste-recycling facilities investment projects to achieve the target. (3) NDRC invites experts to evaluate the action plans following certain rating rules and selects the top-scoring applicants as national urban mining pilot bases. The Ministry of Finance will provide subsidies to support investment of new facilities and equipment as well as collection system. The management procedure of pilot base selection is illustrated in Fig. 7.1.

7.3.2 Progress at Present and Some First Observations

During 2010–2015, NDRC called for five batches of selection; in total 140 recycling industrial parks made application, and only 45 were selected as the national urban mining pilot bases. The 45 selected urban mining pilot base information are shown in Table 7.2. The overview shows that their total planning area is 275.8 Km^2 and total planned annual waste recycling capacity is 68.3 million tons. China in total recycled 210 million tonnes domestic and imported recyclable resources in 2012. The total recycling capacity of 45 urban mining pilot bases can account for one third of this amount, which shows some significance in terms of the scale.

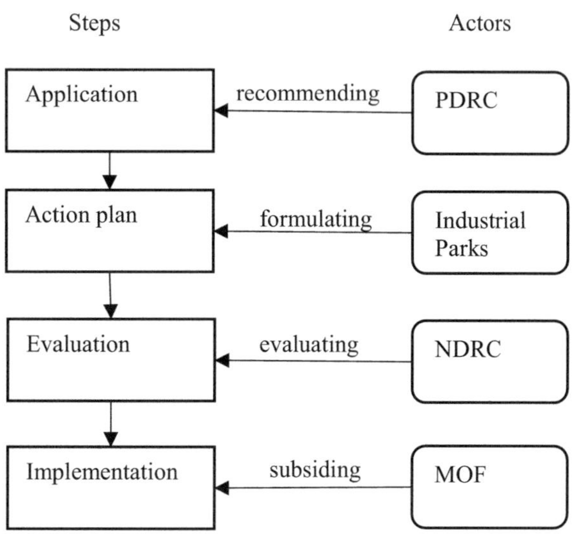

Fig. 7.1 Selection procedure of urban mining pilot bases (*PDRC* Provincial Development Reform Commission, *NDRC* National Development Reform Commission, *MOF* Ministry of Finance). (Authors' contribution)

Table 7.2 Profile of the 45 approved national urban mining pilot bases (Authors' contribution based on own data collection from several sources)

Nr	Name	Planned area (km²)	Total investment (billion RMB)	Recycled in 2012 (million tons)	Planning capacity (million tons)
1	Tianjin Ziya	135	19.8	1.53	3.4
2	Anhui Jieshou	10	10	0.48	0.5
3	Hunan Miluo	18	17.8	1.57	4.0
4	Guangdong Qingyuan	2.7	0.7	0.49	1.2
5	Sichuan Xinan	3.3	0.9	1.03	1.9
6	Qingdao Xintiandi	5.0	1.0	0.37	0.9
7	Zhejiang Ningbo	1.4	5.0	0.4	0.6
8	Shanghai Yanlongji	0.1	0.8	0.49	0.8
9	Guangxi Wuzhou	6.7	15	1.5	1.8
10	Jiangsu Pizhou	4.3	2.0	0.4	0.9
11	Shandong Linyi	1.5	1.1	1.2	3.0
12	Chongqing Yongchuan	2.5	1.5	0.88	0.9
13	Zhejiang Tonglu	8	1.5	1	1.0
14	Hubei Gucheng	10	2.4	1.63	2.8
15	Daliang Eco-park	12	18	0.7	1.9
16	Jiangxi Xinyu	4.2	2.5	2	3.3
17	Hebei Tangshan	1.1	0.7	0.65	1.0
18	Henan Dazhou	10	10.6	2.3	3.5
19	Fujian Huamin	2.3	1.3	0.4	0.3
20	Ningxia Lingwu	6.7	9.0	1	2.1
21	Beijing Lvmeng	0.3	1.0	0.75	1.3
22	Liaoning Donggang	8.2	6.0	0.45	0.9
23	Foshan Yingjia	0.8	2.9	0.58	1.3
24	Anhui Chuzhou	3.8	2.5	0.2	1.0
25	Xinjiang Nanjiang	0.8	1.5	0.4	0.6
26	Shanxi Jitianli	2.7	1.5	0.2	0.5
27	Heilongjiang Dongbu	3.8	0.5	0.37	0.6
28	Hunan Yongxing	4.2	20	0.53	1.4
29	Jilin Gaoxin	2.0	1.3	0.5	0.8
30	Hubei Gelinmei	0.4	4.2	–	1.6
31	Jiangxi Yingtan	10	3.5	–	1.8
32	Jiangsu Rudong	16.6	6.5	–	1.5

(continued)

Table 7.2 (continued)

Nr	Name	Planned area (km^2)	Total investment (billion RMB)	Recycled in 2012 (million tons)	Planning capacity (million tons)
33	Zhejiang Taizhou	4.4	10	–	3.0
34	Hebei Zhonghang	0.7	3.3	–	0.3
35	Sichuan Baohe	3.3	5.0	–	1.7
36	Henan Luoyang	5.9	3.5	–	0.3
37	Guiyang Baiyun	2.2	2.8	–	1.4
38	Fujian Haixi	0.9	3.4	–	2.0
39	Fujian Xiamen	0.4	0.4	–	0.4
40	Shandong Yantai	3.2	2.5	–	5.0
41	Inner Mongolia Baotou	20	26	–	0.8
42	Gansu Lanzhou	14.1	3.2	–	0.8
43	Xinjiang Kelamayi	30	9.0	–	0.5
44	Heilongjiang Haerbin	5.5	5.9	–	2.0
45	Guangxi Yulin	2.0	0.5	–	1.5
	Total	390.8	248.4	24	68.3

Further scanning of these pilot bases shows some disparities between the pilots in terms of the planned area and recycling capacity. This is related with their various historic development foundations and with future goals. The Nr. 1 Tianjin Ziya pilot was a town focused on imported hardware waste dismantling and now targets to become a comprehensive industrial park, including dismantling and recycling of WEEEs, waste vehicles, cables, plastics and cardboards, with a total processing capacity of 3.4 million tons. The planned pilot area also includes the agricultural and residential areas, in total 135 km^2, and a new town. The Nr. 2 Anhui Jieshou pilot is developed from an acid lead battery-recycling aggregation. It is a specifically urban mining industry park focusing on a single waste stream. Due to the national total emission control limitation policy of SO_2, the pilot cannot plan more than 0.5 million tons of capacity.

Figure 7.2 shows the locations of the 45 pilot bases, mainly located at the east coastal and middle region of China. Two factors explain this. First, the industrial development and population are concentrated in the eastern and central regions, leading to a high demand for waste generation and processing as well as for resources. These favour the development of the labour-intensive recycling industry, especially in the provinces of Anhui, Henan and Hebei with large population. Second, coastal areas hold many harbours, through which the waste resources get imported or smuggled in, thus favouring a growth of dismantling and recycling industry in the coastal area. Several urban mining pilots like Tianjin Ziya and Guangdong Qingyuan are developed on such importing and dismantling of wires and cables, WEEEs and waste metals.

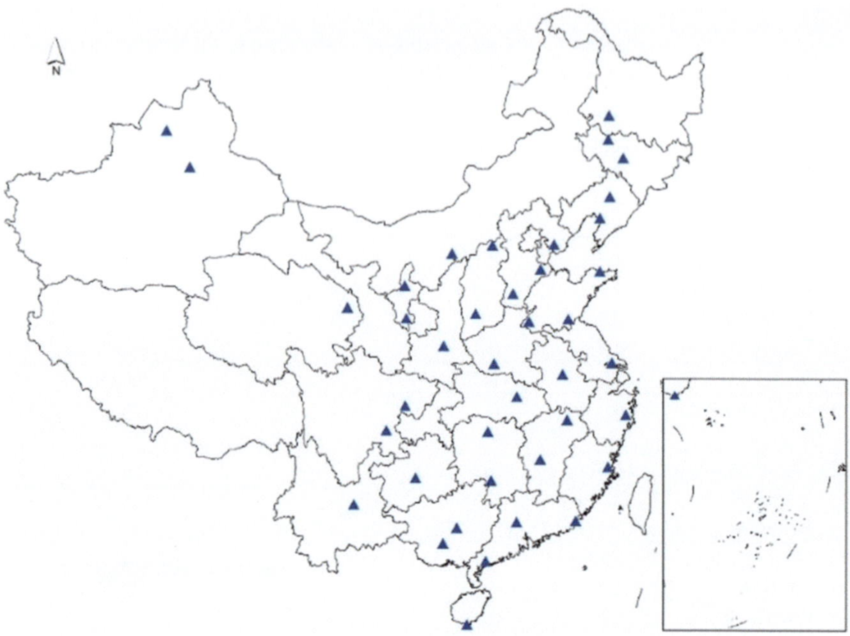

Fig. 7.2 Location of 45 national urban mining pilots bases. (Authors' contribution)

However, the selection of the 45 urban mining pilots was only based on their industry development planning. It has not considered the location distribution issue. Because the inputs for an urban mining pilot are the recyclable waste resources from nearby cities, the waste generation and supply are much related to the local economic development level and the consumption habits of the people. Urban mining pilots cannot be located too nearby each other, to avoid waste supply competition problems. Figure 7.2 shows that there are seven pairs of pilots located nearby each other; their transportation distances are under 200 km. If they focus on similar recycling streams, there must be a potential risk of competition for waste supply. Table 7.3 compares the seven adjacent pilots' similarities and finds that three adjacent pairs of pilots bear such problems.

In summary, some first observations on the 45 urban mining pilots are:

1. The 45 urban mining pilot bases account for one third of the total collected and imported recyclable resources and thus have a high significance.
2. There are many disparities between the pilots in terms of the planned capacity and scale, some are targeting to become a comprehensive recycling industrial park and some are focused on specific waste streams.
3. The selection of the urban mining pilots has neglected the geographical distribution issue, resulting in seven pairs of pilots that are located within 200 km from each other, and some of them have a similar industry planning, thus leading to a high risk of waste resource competition in the future. The overall management of these urban mining pilots should coordinate the capacity planning among the pilots.

Table 7.3 Mutual distances of seven pairs of pilots are less than 200 km (Authors' contribution)

Adjacent pilots	Distance (km)	Similarity of their recycling industry
Xiamen and Quanzhou	100	Same recycling industry but different waste resources, middle similarity
Yongchuan and Neijiang	112	Both are comprehensive recycling industry parks, high similarity
Jingmen and Xiangyang	124	Different waste streams focused on, low similarity
Beijing, Tianjin, Tangshan	125	Three are comprehensive recycling industry, high similarity
Nantong and Shanghai	128	One comprehensive and the other is specific glass recycling park, low similarity
Taizhou and Ningbo	174	Middle similarity
Linyi and Xuzhou	198	Middle similarity

7.4 The Policy Analysis

This section presents a policy analysis of the UMPB programme. We firstly review other relevant policies and find the policy evolution path and then profile its policy network which is featured with a multi-ministries cross-management system. Finally, we apply governance contextual analysis tool to find the supportiveness of the current governance structure for the programme.

7.4.1 Policy Evolution of the Urban Mining Programme

The urban mining programme is an important initiative for developing circular economy in China. Reviewing circular economy relevant policy suggests that the urban mining initiative is not isolated but was built upon other policies. It also shows that the driving forces of waste-recycling management in China are moving from environmental and recycling angles to resource strategy concerns.

The circular economy concept was first officially introduced by the Ministry of Environmental Protection (MEP) in 2002 but soon included in the profile of NDRC authorities. Table 7.4 lists all circular economy-relevant policies since then.

In 2004, MEP initiated a circle zone management for the assembling area of imported waste-recycling workshops (mainly wastes of electric wire, cable, machinery and equipment). The objective was to improve the environmental management in the area. Nineteen pilots were selected; instruments to monitor imported waste licences and total control as well as environmental technology planning were introduced to improve in total 627 enterprises.

In 2005 and 2007, NDRC with other five ministries jointly initiated two batches of circular economy pilots, in total supporting 178 pilots to practise waste reduction, recycling and recovery projects (Jiao and Boons 2014). Among these pilots, some

Table 7.4 Relevant national circular economy policies and programmes (Authors' contribution)

National programmes	Year started	Ministry in charge	Policy focus	Results so far
Eco-industrial park	2003	MEP	Recognise eco-industrial parks	26 were accomplished; 59 are in application procedure
Circle zone management initiative	2004	MEP	Improve environmental management of waste-recycling assembling area	19 pilots were selected
Circular economy demonstration programme	2005 2007	NDRC	Key industries, industrial parks, province and cities	178 demonstration projects were accomplished
Urban mining pilot base programme	2010	NDRC	Supports 50 pilots bases	45 pilots selected
Industrial park circularised development programme	2012	NDRC	Supports 100 industrial parks to implement circular economy projects	67 industrial parks were selected
Circular cities demonstration programme	2013	NDRC	Supports 100 cities	First batches, 40 cities were selected

waste-recycling projects were included in the first batch such as renewable resource collection and recycling, waste metals recovery, WEEE and remanufacturing. The second batches list several renewable resource-recycling industrial parks, e.g. Tianjin Ziya and Anhui Jieshou, which were naturally selected as urban mining pilot bases in 2010.

The above two initiatives addressed different issues of recycling. The circular economy pilots explored various circular economy models at wide-ranging fields and industries. A total of 178 pilots included enterprises, industrial parks, provinces and cities, of which nine circular economy pilot parks were further selected as pilot in urban mining pilot base programme in 2010. The circle zone management initiative targeted on environmental issues of the recycling companies. There are only 19 pilots, but ten of which were also selected as urban mining pilot bases later. Sixteen UMPBs enjoyed the circle zone management policy or circular economy pilots before they were listed as urban mining pilot bases (see Table 7.5).

The urban mining programme has the dual objective of industrial development and reducing environmental and resources constrains, but its resource strategy is more obvious as it sets an annual capacity threshold of 0.3 million tons for the applicants. Figure 7.3 illustrates the policy evolution path of urban mining policy. It shows the driving forces of waste-recycling management in China moved from environmental and recycling angles to resource strategies at UMPB programme and then to an integration of three aspects in 2013 when NDRC initiated 100 circular pilot cities programme to promote comprehensive circular economy development of industrial, agricultural and society sectors in cities (NDRC 2013). Till then, China is moving forward to a more comprehensive urban symbiosis stage.

Table 7.5 Some urban mining bases are developed from the circular economy pilot and circle zone management pilot programmes (Authors' contribution)

Nr.	Urban mining pilot bases	Urban mining pilot base approving date	Circular economy pilots approving date	Circle zone pilots approving date
1	Tianjin Ziya	2010	2007	2007
2	Anhui Jieshou	2010	2007	–
3	Hunan Miluo	2010	2005	–
4	Guangdong Huayuan	2010	2005	–
7	Zhejiang Ningbo	2010	2007	–
9	Guangxi Wuzhou	2011	–	2010
15	Dalian Eco-park	2011	–	2012
18	Henan Dazhou	2011	2007	–
22	Liaoning Donggang	2011	–	2010
28	Hunan Yongxin	2012	2007	–
31	Jiangxi Yingtan	2013	2005	2009
32	Jiangsu Rudong	2013	–	2011
33	Zhejiang Taizhou	2013	–	2008
40	Shandong Yantai	2014	–	2007
41	Inner Mongolia Baotou	2014	2005	–
45	Guangxi Yulin	2014	–	2008

7.4.2 The Governance of Implementing the UMPB Programme

The urban mining policy has evolved from the previous relevant circular economy policies but is also supported by a more comprehensive policy framework. Together they define the policy network (see Fig. 7.4). The term policy network is here not confined to the relevant actors but also to the policies, institutions and resources that form the context of the policy implementation. This is also labelled as the "governance" context (Hans Bressers and Kuks 2004).

Firstly, at the regulation level, the Circular Economy Promotion Law promulgated in 2009, provides a regulation basis for all circular economy planning and pilot programmes including the urban mining programme (Su et al. 2013). Secondly, at the planning level, the 12th national Five-Year Plan (FYP) sets the target to increase resource productivity by 15%. This is the first time for China that a FYP sets target on resources and recycling. Following that, the NDRC has drawn up a circular economy development strategy and action plan, which lists major tasks such as the top 10 circular economy (CE) pilot projects, 100 CE pilot cities and 1000 CE pilot enterprises and parks. The urban mining pilot base programme is emphasised in the plan. There are other special plans that also support and embody the urban mining programme (see Table 7.6).

Fig. 7.3 Policy evolution path of urban mining policy. (Authors' contribution)

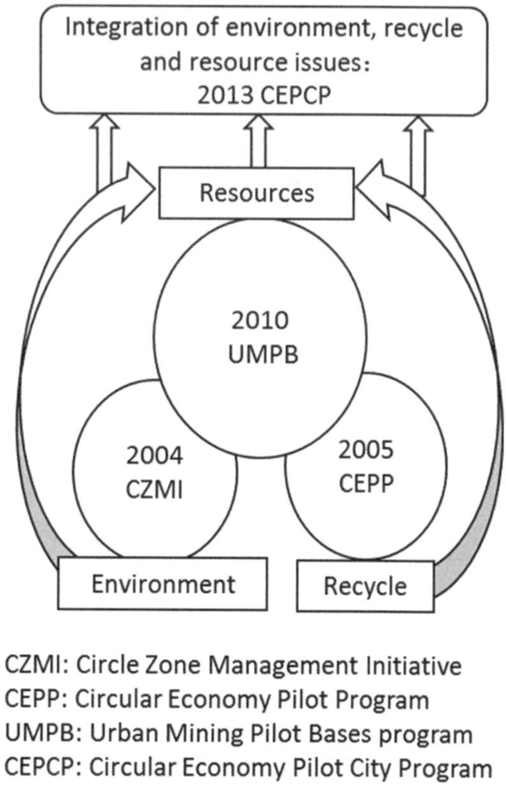

Lastly, the urban mining programme also receives financial support. The Ministry of Finance established a circular economy fund, to support projects in six sectors, which include urban mining pilot base programme, food waste management, industrial park circular retrofit, remanufacture, cleaner production technology promotion and circular economy infrastructure. The fund provides 10% of the total investment ratified in the urban mining pilot base action plan as subsidy. About 4 billion RMB has been ensured for the urban mining pilot bases. Averagely, every pilot base can receive about 0.1 billion RMB for the projects construction.

However, while the current urban mining policy framework looks sufficiently supportive in terms of policy and planning context and resource allocation, there are still some issues to be improved.

The single urban mining pilot base programme covers seven major types of waste. It tries to establish a comprehensive platform for waste recycling. Comparing with the EU waste management legal framework (see Table 7.7), the EU sometimes deploys specific directives to regulate specific waste management streams. But the target of China's urban mining policy is not the waste itself but industrial park which provides a platform for large-scale and high-value recycling production. Therefore, it also needs other policy mechanisms such as pollution control, recycling licences, industry access, waste collection system, financing and industry

Fig. 7.4 Legal framework of UMBP programme. (Authors' contribution)

chain cultivation. Because different wastes are under different ministries' jurisdictions, this leads to a complex urban mining policy network consisting of multiple ministries and multiple waste streams (see Figs. 7.4 and 7.5).

In this multi-ministerial cross-management network, policy conflicts occur in many cases. The collection, recycling and pollution control parts are under different ministries' jurisdictions. Some planned recycling facilities can be hindered by licence applications from other departments and even undermined due to short supply of waste resources.

The absence of a specific waste management law is another shortage in the current governance context of implementing the urban mining policy. Particularly waste collection is not supported by regulation but only relies on free market mechanisms. This leads to several problems. Waste streams can be transported more than 500 km to the highest price buyer while nearby plants are in short supply; thus eco-efficiency is not assured.

Currently the urban mining pilot base programme policy effects are not coordinated with the WEEE policy. Since 2012, the Ministry of Environmental Protection initiated a WEEE-recycling and management programme, a fair WEEE fund and management system has been established, and 106 WEEE-dismantling enterprises are listed to receive subsidies. However, only few enterprises are located in the current 45 urban mining pilot bases; therefore the two policies have little mutually supportive influence to one another.

Table 7.6 Several 12th FYP special plans reinforce the urban mining pilot base programme (Authors' compilation from various sources)

	Ministry in charge	Special plans	Main tasks listed in the plans relevant to urban mining and resource recycling
2011	State council	12th FYP action plan for energy saving and emission reduction	100 resource comprehensive utilisation bases
			80 waste collection pilot cities
			50 urban mining pilot bases
			5 remanufactures' assemble area
			100 food waste management pilot projects
2011	NDRC	Guidelines for comprehensive resource utilisation of 12th FYP	Increase renewable resource collection rate to 70%
			Increase secondary copper, aluminium and lead production to 40%, 30% and 40% of the total production, respectively
2012	Ministry of Science and Technology	12th FYP special plan of waste-recycling technology projects	Priorities filed include resource recycling technologies of metals, WEEEs and polymer and electric machine remanufacturing
2012	State council	12th FYP special plan of energy saving and environmental industry development	Support 50 urban mining pilot bases; support waste collection, recycling industrial chain, environmental pollution remediation infrastructure and platform, scaled and high-value resource recycling
2012	State council	12th FYP special plan of national strategic emerging industries development	Support some urban mining pilot bases, and improve the recycling technology and equipment manufacture level of waste metals, rubble, tire, battery, etc.
2012	State council	12th FYP special plan of energy saving and emission reduction	Implement key projects of resource comprehensive utilisation, waste collection, urban mining, remanufacture, food waste management, industrial park circular reform, technology spreading
2013	State council	Circular economy development strategy and short-term action plan	Top 10 pilot projects: waste collection pilots, 50 urban mining pilot bases
			100 CE pilot cities
			1000 CE pilot business/industrial parks

Table 7.7 Comparison between and EU waste management directives (Authors' contribution)

Waste stream covered by china UMPB programme	Corresponding EU directives
Electronic equipment	Directive on WEEE (2002/96/EC)
Cables	Directive on WEEE (2002/96/EC)
Communication facilities	Directive on WEEE (2002/96/EC)
Household appliances	Directive on WEEE (2002/96/EC)
Electronic products	Directive on WEEE (2002/96/EC)
Packaging and scraps	Directive on packaging and packaging waste (1994/62/EC)
End-of-life vehicle	Directive on end-of-life vehicle (2000/53/EC)

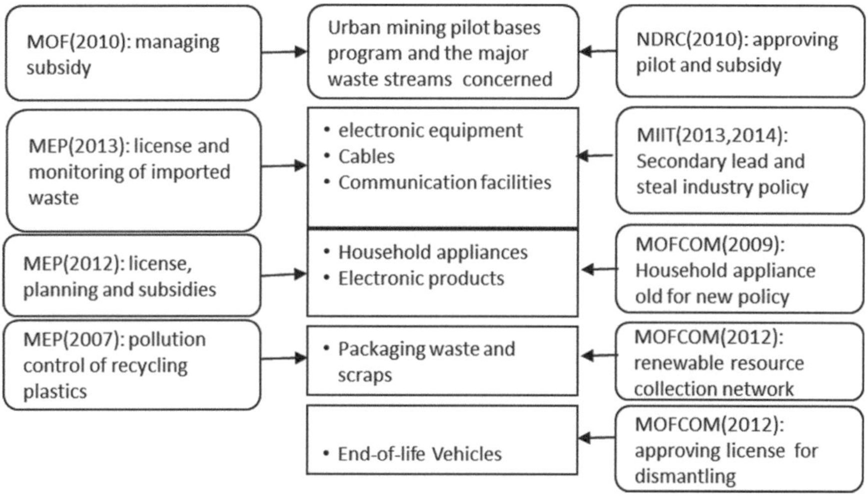

Fig. 7.5 A complex urban mining policy network consisting of multiple ministries' cross-management (*MOF* Ministry of Finance, *MEP* Ministry of Environmental Protection, *NDRC* National Development and Reform Commission, *MIIT* Ministry of Industry and Information Technology, *MOFCOM* Ministry of Commerce). (Authors' contribution)

All in all, the current urban mining governance context only provides a platform for the urban mining recycling activities, some policy conflicts still occur and special regulations are absent, so inter-ministerial coordination and policy integration are needed to develop an overall institutional design for urban mining in China. Consequently, specific recycling-oriented regulations are in need to assure sufficient waste supply for the urban mining industry development. Lastly, coordination between ministries and policy integration is needed to implement the urban mining pilot base policy goal of resolving resource bottleneck constraints.

7.4.3 Analysing the Supportiveness of the Governance Context

Above we have shown that the urban mining policy in fact originates from many policies and with many governmental stakeholders. It attempts to harmonise those in an overall platform. Hans Bressers and Kuks (2004) state that under these conditions, two criteria determine the degree to which the governance context, including the network context, is supportive for the implementation. The criteria are defined by the questions that they pose to the five dimensions of governance (administrative levels and geographical scales, actors and networks, problem perspectives and goal ambitions, strategies and instruments and responsibilities and resources for implementation):

1. *Extent*: are all elements in the five dimensions that are relevant for the policy or project that is focused on taken into account? Or are essential ones lacking?
2. *Coherence*: are the elements in the dimensions of governance reinforcing rather than contradicting each other?

Sufficient extent and coherence together form genuine policy integration. Just having sufficient extent, being complete, is not enough. However, lack of extent can be one of the sources of lack of coherence.

In later theoretical development and empirical studies, two extra criteria were added:

3. *Flexibility*: are multiple roads to the goals, depending on opportunities and threats as they arise, permitted and supported?
4. *Intensity*: how strongly do the elements in the dimensions of governance urge changes in the status quo or in current developments?

For each of the five dimensions of governance, the four criteria mentioned above can be applied, which forms a matrix. This matrix forms the core of the *Governance Assessment Tool* (GAT) (Hans Bressers et al. 2016). Together, the questions in each cell shed light on the degree of supportiveness or restrictiveness of the governance context towards the implementation of policies and projects. Note that not the implementation and success of the programme itself is evaluated, but the degree to with the governance context is supportive for such success.

While the early assessment of the governance context as described in the sections above does not provide detailed information on all the 20 separate cells, it is possible to systematically analyse the governance context for the implementation of the urban mining pilot base programme with the criteria of the governance assessment tool.

Extent: High Level of Involvement, but Some Elements Are Still Lacking
The selection process involved both upper and lower levels in a sort of top-down bottom-up interaction. Also the potential bases themselves were involved. This selection process has however not resulted in an optimal geographical spread of pilot sites. Among the problem perspectives, obviously such geographical scope has been missing.

Absence of a specific waste management law is another shortage in the current policy governance context. Particularly waste collection is not supported by regulation but only relies on free market mechanisms. This leads to several problems, as waste can be transported more than 500 km to the highest price buyer while nearby plants are in short supply; thus eco-efficiency is not assured, and extra unnecessary pollution is likely. While reducing these was an important purpose of the national programme, these shortages affect the core of the policy.

Coherence: Both Reinforcing and Contradicting Forces
On the positive side, the urban mining pilot base programme is very well embedded in the general circular economy policies, laws and plans of China. Equally positive is that the development of those policies and plans goes consistently into the

direction of more comprehensive strategies, for instance, with the recent 100 circular city pilot programmes.

On the other hand, the broad scope of the programme, involving various waste streams and stages of processing, also creates coherence issues. It requires a multi-ministerial cross-management network while the collection, recycling and pollution control parts are under different ministries' jurisdictions. They are not always geared towards mutual cooperation. As a consequence, in many cases policy conflicts occur.

While the geographical distribution of the pilots is not optimised, it is possible that in some cases completion for waste streams occurs, leading to transports and pollution hazards.

While the WEEE's policy is not coordinated with the urban mining pilot base programme, only few enterprises are located in the current 45 urban mining pilot bases. Thus the potential for reinforcing the two programmes is underused.

Flexibility: Test Is Yet to Come

The criterion of flexibility is an important governance asset in situations in which the field develops in a dynamic way, and thus adaptive responses to these changes are required to keep the original goals feasible. It can be predicted that such developments are bound to occur in the future. One might think of technological developments in waste separation and processing or of developments in supply of waste stream and demands for materials due to technological and economic forces like the gradual transition towards a service-oriented economy. As the implementation of the programme is still under development, the test whether it is capable of allowing and supporting a sufficient degree of flexibility to the bases to cope with such changes while keeping the objectives up is yet to come.

Intensity: High Level of Policy Support

While the urban mining pilot base programme is one of the spearheads of the circular economy policy turn that China is pursuing, even supported by the 12th FYP, it has a high level of policy support that makes it difficult to ignore. The sheer size of the effort to create a nationwide innovative system testifies this. Whether the level of finance (10% subsidy) is sufficient to enable the pilots to flourish remains to be seen. Probably solving the issues mentioned under extent and coherence is as important as that and will also require continuously strong high-level policy support.

Highlights

The systematic analysis of the governance context above confirms the picture of the previous section. While intensity of the programme is quite high, the institutional organisation created a medium level of extent and coherence. Whether flexibility will prove sufficient is too early to tell. Actually this picture is not uncommon for an innovative programme. By its nature it has to obtain a position among many existing and more established policies and government organisations. Only an open eye for its remaining inconsistencies and a continuation of its high-level support can make the situation develop into an even more supportive governance context, for instance, by taking the measure mentioned at the end of the previous section.

An open mind includes a desire to learn from previous examples. Therefore, the next section will compare the Chinese programme with a programme that Japan started already in the late 1990s.

7.5 Comparison with the Eco-town Programme in Japan

Facing a similar waste and resources problem as China did later when developing economically, Japan initiated its eco-town programme in 1997 with two aims: to extend landfill site life and to revitalise local industry. Local governments formulated eco-town plans and submitted them to the ministries for approval and endorsement. Ministries provided grants to local authorities to execute the town planning, community recycling and outreach activities and subsidies to private companies to invest in the recycling projects (Van Berkel et al. 2009). During 10 years of operation, 26 eco-towns are endorsed, and 205 projects were invested in and started in operation. The eco-towns are classified into three types: promotion of environmental industries, treatment of wastes and community development. The projects are mainly focused on plastic recycling as the largest waste stream and food and electronic waste recycling (Ohnishi et al. 2012). The programme has been proven successful. It not only contributed to diversification and sophistication of recycling technologies such as metal recovery and high-grade recycling options for plastics but also had a broader impact as eco-towns are regarded as industrial recycling clusters with extensive cooperation among different companies in Japan's Second Fundamental Plan for Establishing a Sound Material-Cycle Society (OECD 2011). Table 7.8 summaries comparative information between Japan's eco-towns programme and the urban mining base programme in China.

Although the eco-town programme targets at the city level, while the urban mining programme is targeting at industrial park level, it is still worthwhile to compare the pair, as both programmes support the same category of recycling projects: recycling waste from urban life. This makes them sharing a similarity in terms of urban symbiosis. The total capacity of eco-towns' recycling projects is almost 2 million tons per year, and China's urban mining bases' planned capacity is 66 million tons per year. Other similarities include that both programmes share the same objective to bloom the recycling industry development and share the same national policy background of striving for a circular society.

Still, both programmes also present diversities of the urban symbiosis, these include:

1. Recycling projects in the eco-town programme are mainly focused on the municipal waste recycling such as plastics, food waste and electric waste, while China's urban mining base programme supports the recycling plants of electronic equipment and products, cables, vehicles and tires, packaging, etc. Municipally gener-

Table 7.8 Comparison between Japan's eco-town programme and China's urban mining programme

	Japan eco-town programme	China urban mining programme
Time	1997–2006	2010–2015
Policy objectives	Stimulating new industrial development	Stimulating new industrial development
	Addressing waste management issues	Help relief environmental and resource bottleneck
Ministries in charge	Ministry of Environment (MoE),	National Development and Reform Commission (NDRC)
	Ministry of Economy	
	Trade and Industry (METI)	
Legal bases	Basic Law for Establishing the Recycling Society	Circular Economy Promotion Law as general law
	National planning	12th FYP and specific plans
	Specific laws (2002–2003)	Absence of specific laws
Result	26 eco-towns endorsed	Target to support 50 bases; 45 are already recognised and 5 more to come
	205 projects (170 were recycling and recovery projects in operation, of which 61 received subsidies)	
Investment and subsidies	Grants to local government for planning execution activities, 50% of project costs in the range of 3–5 million JPY/year (30–50,000 USD/year) for a 3–5-year period.	No grants to local authorities
	Subsidy to companies for recycling plants at averagely 36% of the investment (total investment 1.65 billion USD), total subsidy 59 billion JPY (approximately 590 million USD) spent	Total projects investment is 192.3 billion RMB (approximately 32 billion USD). Subsidy is 10% of total investment in every urban mining base, 4 billion RBM (approximately 666 million USD) ratified for 45 bases
Project types	Software projects of the community and outreach activates	Software projects include waste collection system and waste information platform projects
	Plants' projects include plastic-recycling projects, as the largest group, and food and electronic waste-recycling projects	Other major projects include recycling plants of waste metals, electronics, vehicles and tires, plastics etc.; no food waste stream
Effects and impacts	It contributed to recycling technology development	It is still early to do programme evaluation, but some effects already occur. The programme helps to upgrade recycling technology and equipment and integrate waste collection systems and the industrial chains in some bases. Assembling and upscaling waste recycling also generate environment impacts
	Eco-towns are recognised as industrial recycling clusters with extensive cooperation among different companies in the Second Fundamental Plan for Establishing a Sound Material-Cycle Society	

Eco-town programme information is adapted from Fujita (2008), GEF (2005), Sato et al. (2004) and Van Berkel et al. (2009)

ated food wastes are not included but covered by another national initiative. This is related to the different policy objectives between two programmes. The eco-town programme is to extend the service life of landfilling site and to cultivate local industry, while the urban mining programme is to promote the recycling industry development for resource and environmental concerns.

2. The slight difference of policy background and policy objectives also causes that the scope of urban symbiosis in both programmes is different. Recycling projects in the eco-town programme are mainly processing waste from local and nearby cities. The recycling boundaries of China's urban mining bases can extend to provinces at 500 km away. Moreover, some recycling plants in Japan are facing a shortage of waste supply because of exporting to Chinese plants. The scope of urban symbiosis in China has a larger scope than Japan. But eco-efficiency of both large- and small-scope urban symbiosis needs further study.
3. Several factors contribute to the success of the eco-town programme, but the recycling-oriented legislation is conceived as an most important one to assure sufficient supply of waste to the recycling plants (Van Berkel et al. 2009). The Waste Management Law (2003) sets aims and objectives for waste management. The Law for Promotion of Effective Utilisation of Resources (2001) designated key products and industries for resource saving. Besides, the Law for Promotion of Sorting, Collection and Recycling of Containers and Packaging (2000), the Law for Recycling of Specific Kinds of Home Appliances (2001), the Construction Materials Recycling Act (2002), the Food Recycling Law (2003) and the Domestic Automobile Recycling Law (2003) all set very specific recycling goals (Morioka et al. 2005). These recycling-oriented legislations are still absent in China's legal framework.

In summary, the similarity between the eco-town programme and the urban mining programme is based on their joint focus on urban symbiosis. The differences between both programmes indicate their different background and different policy objectives. Industry development for resources is the core feature of China's urban mining programme, and waste management is attached to the eco-town programme. This leads to different results of small scope urban symbiosis in Japan and large scope urban symbiosis in China. And it is too early to conclude which one is more eco-efficient.

The eco-town programme is a proven success, contributing to recycling technology diversification and sophistication while being recognised in the later Japan Second Fundamental Plan for Establishing a Sound Material-Cycle Society. While it is too early to evaluate the urban mining programme, already some obvious effects are appearing. Firstly, it helps technology and equipment upgrading. Jieshou, a specialised waste lead acid battery-recycling park, discarded hand-dismantling equipment and deploys the most advanced automatic production line, vastly improving the efficiency and reducing environmental and health risks. Secondly, it promotes extending of industrial recycling chains. Many bases planned facilities for new production by using recycled materials. Also some bases expand the waste resource collection systems to ensure waste supply for the recycling plants. All these improvements are attributed to the urban mining programme that provides a platform and subsidies for massive resource recycling. Last but not least, the environment effect is obvious when the technology and equipment are upgraded and when waste-recycling quantities are increased.

7.6 Conclusion

Urban symbiosis taking recycling as main activity links efficient material flows between production and consumption the industry sector and urban sector. China's urban mining pilot base programme promotes the recycling industry developing towards large-scale, advanced technology and high-value recycling practise to help in relieving environmental and resource bottleneck constraints. The current 45 pilot bases with planned 6.6 billion tons per year capacity indicate that a massive urban symbiosis effort is taking place in China.

This paper provides a preliminary review of the urban mining pilot base profile by analysing its policy evolution pass and the policies in which it is located. It finds that China's urban mining base programme is developed from past circular economy policies, including the circular economy pilot programme and circular zone management programme. It shows that the driving forces of waste recycling management in China are moving from an environmental and recycling angle to resource strategies. The programme attains legal assurance from the Circular Economy Promotion Law and support from the national 12th FYP and specific plans, as well as subsidies from the circular economy fund. But this does not yet create a perfect governance context for its implementation. On the contrary, a multi-ministerial cross-management network is currently implementing the urban mining pilot base programme. In this network, incoherencies create, policy conflicts occur, and recycling-oriented legislation is absent, thus becoming main barriers for the urban mining programme and especially its waste collection requirements.

Comparing with the eco-town programme in Japan, the urban mining programme shares the partial policy objective to promote industry development. But waste management and environmental amenity is another driving force for the eco-town programme, while optimising the resource strategy is for the urban mining programme. This leads to the differently focused recycling projects and recycling area boundaries between the two programmes. Therefore, China's urban mining programme is regarded as a large scope urban symbiosis programme and Japan's eco-town as a local scope urban symbiosis. Such difference is attributed to the different problems they attack and the slightly different policy objectives under different economic development phases. The resulting eco-efficiency of both forms is a future study subject.

It is too early to evaluate the urban mining programme; still some obvious effects are already appearing. The recycling technology and equipment improved, industrial recycling chains extended up to collection part and down to production part, and environmental effects are observed. As it stimulates massive recycling activities, the urban mining pilot programme will help with environmental and resource constraints. However more inter-ministerial coordination and policy integration are needed to develop the governance context into an institutional top design for a sustainable urban mining in China.

References

Bressers H, Kuks S (2004) What does governance mean? From conception to elaboration. In: Bressers H, Rosenbaum W (eds) Achieving sustainable development. The challenge of governance across social scale. Praeger, Westport/London, pp 65–88

Bressers H, Bressers N, Kuks S, Larrue C (2016) The governance assessment tool and its use. In: Bressers H, Bressers N, Larrue C (eds) Governance for drought resilience: land and water drought management in Europe. Springer, Cham, pp 45–65

Chertow MR (2000) Industrial symbiosis: literature and taxonomy. Annu Rev Energy Environ 25(1):313

Chertow MR (2007) "Uncovering" industrial symbiosis. J Ind Ecol 11(1):11–30. https://doi.org/10.1162/jiec.2007.1110

Dong H, Ohnishi S, Fujita T, Geng Y, Fujii M, Dong L (2014) Achieving carbon emission reduction through industrial & urban symbiosis: a case of Kawasaki. Energy 64(0):277–286. https://doi.org/10.1016/j.energy.2013.11.005

Fujita T (2008) Eco-town projects/environmental industries in progress: environment conscious type of town building. Ministry of Economy, Trade and Industry, Tokyo

GEF (2005) Eco-towns in Japan: implications and lessons for developing countries and cities. Retrieved from Osaka, Japan

Geng Y, Tsuyoshi F, Chen X (2010) Evaluation of innovative municipal solid waste management through urban symbiosis: a case study of Kawasaki. J Clean Prod 18(10–11):993–1000. https://doi.org/10.1016/j.jclepro.2010.03.003

Golev A, Corder GD, Giurco DP (2014) Industrial symbiosis in Gladstone: a decade of progress and future development. J Clean Prod 84(0):421–429. https://doi.org/10.1016/j.jclepro.2013.06.054

Graedel TE (2011) UNEP recycling rates of metals – a status report, a report of the working group on the global metal flows to the international resource panel. https://doi.org/ISBN 978-92-807-3161-3

Jacobs J (1969) The economy of cities. Random House, New York

Jiao W, Boons F (2014) Toward a research agenda for policy intervention and facilitation to enhance industrial symbiosis based on a comprehensive literature review. J Clean Prod 67(0):14–25. https://doi.org/10.1016/j.jclepro.2013.12.050

Johansson N, Krook J, Eklund M, Berglund B (2013) An integrated review of concepts and initiatives for mining the technosphere: towards a new taxonomy. J Clean Prod 55(0):35–44. https://doi.org/10.1016/j.jclepro.2012.04.007

Krook J, Baas L (2013) Getting serious about mining the technosphere: a review of recent landfill mining and urban mining research. J Clean Prod 55:1–9. https://doi.org/10.1016/j.jclepro.2013.04.043

Krook J, Eklund M, Carlsson A, Frändegård P, Svensson N (2010) Urban mining: prospecting for metals in the invisible city. Delft University of Technology, Delft

Minter A (2013) How Beijing and the rest of China – recycles plastic. Retrieved from https://www.scientificamerican.com/article/china-recycles-plastic/

MLR (2016) China mineral resources. Retrieved from Beijing: http://www.mlr.gov.cn/sjpd/zybg/

Morioka T, Tsunemi K, Yamamoto Y, Yabar H, Yoshida N (2005) Eco-efficiency of advanced loop-closing systems for vehicles and household appliances in Hyogo eco-town. J Ind Ecol 9(4):205–221. https://doi.org/10.1162/108819805775247909

NDRC (2010) Notification of establishing national urban mining pilot bases program (in Chinese). Retrieved from http://www.gov.cn/zwgk/2010-05/27/content_1614890.htm
NDRC (2013) Notification of initiating circular economy pilot cities program (in Chinese). Beijing
OECD (2011) Setting and using target sustainable materials management opportunities and challenges. Retrieved from Paris
Ohnishi S, Fujita T, Chen X, Fujii M (2012) Econometric analysis of the performance of recycling projects in Japanese eco-towns. J Clean Prod 33(0):217–225. https://doi.org/10.1016/j.jclepro.2012.03.027
Sato M, Ushiro Y, Matsunga H (2004) Categorisation of eco-town projects in Japan. Paper presented at the international symposium on green technology for resources and materials recycling, Seoul, Korea
Shi H, Chertow M, Song Y (2010) Developing country experience with eco-industrial parks: a case study of the Tianjin economic-technological development area in China. J Clean Prod 18(3):191–199
Song Q, Li J (2014) Environmental effects of heavy metals derived from the e-waste recycling activities in China: a systematic review. Waste Manag 34(12):2587–2594. https://doi.org/10.1016/j.wasman.2014.08.012
Su B, Heshmati A, Geng Y, Yu X (2013) A review of the circular economy in China: moving from rhetoric to implementation. J Clean Prod 42(0):215–227. https://doi.org/10.1016/j.jclepro.2012.11.020
UNEP (2010) Metal stocks in society: scientific synthesis
Van Berkel R (2004) Industrial symbiosis in Australia: an update on some developments and research initiatives. Paper presented at the the industrial symbiosis research symposium at Yale: advancing the study of industry and environment, New Haven, CT
Van Berkel R, Fujita T, Hashimoto S, Geng Y (2009) Industrial and urban symbiosis in Japan: analysis of the eco-town program 1997–2006. J Environ Manag 90(3):1544–1556
Wen Z, Zhang C, Ji X, Xue Y (2015) Urban mining's potential to relieve China's coming resource crisis. J Ind Ecol 19(6):1091–1102. https://doi.org/10.1111/jiec.12271
Xing GH, Chan JKY, Leung AOW, Wu SC, Wong MH (2009) Environmental impact and human exposure to PCBs in Guiyu, an electronic waste recycling site in China. Environ Int 35(1):76–82. https://doi.org/10.1016/j.envint.2008.07.025
Xu X, Yang H, Chen A, Zhou Y, Wu K, Liu J et al (2012) Birth outcomes related to informal e-waste recycling in Guiyu, China. Reprod Toxicol 33(1):94–98. https://doi.org/10.1016/j.reprotox.2011.12.006
Xue Y, Wen Z, Ji X, Bressers HTA, Zhang C (2017) Location optimization of urban mining facilities with maximal covering model in GIS: a case of China. J Ind Ecol 21(4):913–923. https://doi.org/10.1111/jiec.12467
Zhang Y (2010) Development of waste plastic recycling industry in Gengche Town, Suqian of Jiangsu Province (in Chinese). Retrieved from https://wenku.baidu.com/view/1779ae176c175f0e7cd137ff.html
Zhao M (2011) Recycling industry accounts for 40% of GDP contribution in Yujiang County. Retrieved from http://jiangxi.jxnews.com.cn/system/2011/08/01/011733028.shtml

Chapter 8
Incorporating Circular Sustainability Principles in DKI Jakarta: Lessons Learned from Dutch Business Schools Management

Juli Nurdiana, María-Laura Franco-García, and Sharon Hophmayer-Tokich

Abstract This study aims to set future direction of sustainability pathway(s) to incorporate circular economy within Indonesian high education systems and to "reinvent" the university role whilst shaping future leaders. For this purpose, this study intended to develop the definition of circular sustainability high education. Some of the Indonesian universities have already adopted the sustainability principles, putting the circularity concept at the heart of their education system. Nevertheless, by comparing them with some of the international experiences, in specific those of the Dutch universities, it was expected to identify relevant opportunities for the Indonesian universities to enhance their contribution on the circular sustainability fields. In consequence, the approach of this study was to analyse the managerial practices in order to integrate circular economy at Dutch universities (e.g. Business Schools), which represented the analytical framework for the Indonesian case. In terms of the research question driving this work, it corresponded to: "What are those transferable Dutch practises of circular economy towards sustainability transition to Indonesian Business Schools in DKI Jakarta?" From the methodological viewpoint, this study presented an exploratory and comparative design. As result of this research, it was observed that, among the cases analysed, there is a large difference in the management approach, particularly on how the university leaders demonstrate their contribution to sustainability. This latter requires, consequently, a strategic and systemic approach to measure the circular sustainability through performance indicators to assess progress which are spread out across four criteria. Some of the conclusions showed that there is currently no fixed panacea on how to integrate the concept of circular sustainability high education into the universities culture. Even further, the result of this

J. Nurdiana (✉)
Faculty of Engineering, Mulawarman University, Samarinda, Indonesia

M.-L. Franco-García
University of Twente, Enschede, The Netherlands
e-mail: m.l.francogarcia@utwente.nl

S. Hophmayer-Tokich
CSTM Department of Governance and Technology for Sustainability, University of Twente, Enschede, The Netherlands
e-mail: s.hophmayer-tokich@utwente.nl

M.-L. Franco-García et al. (eds.), *Towards Zero Waste*, Greening of Industry Networks Studies 6, https://doi.org/10.1007/978-3-319-92931-6_8

study could proffer a guide towards a shift in circular sustainability and be used for detailed further application to depict the stand for Indonesian Business Schools.

Keywords Circular economy · Transition · Sustainable university · Indonesian management programme

8.1 Introduction

It is growingly recognised that we must fundamentally rethink our current liner economic model due to the increased pressure on the resources. The idea of shifting to a more circular model to relieve the escalating pressures on our resources – energy, materials and water – is influencing business leaders, global companies and institutions. Business leaders acknowledge that in order to secure future generations, a reconciliation of industrial model and joint plan actions is required and more and more business leaders have lent their endorsement to the principle of circular economy, particularly on their role and of what they count as success (Gitsham et al. 2013). As such, many large corporations have piloted their business models based on the extended life cycle thinking. Unilever, Coca-Cola company, BP, Shell, Heineken, IMC Pan Asia Alliance group and Accenture are some of the companies that integrate circular economy principles in their management and business (Gitsham et al. 2013). One possibility of scaling this trend up is through introducing a circular economy by designing the drivers, for example, by embarking on the role education can play in shaping future business leaders. This can be done by exploring how an educational institution can deliver effective education that will have a significant influence on future leaders in the move towards making the transition to a circular economy. A question is also raised on the institution platform to connect business leaders with circularity.

Universities have distinctive organisational cultures that value and promote learning and thus can play a vital role in processes of societal transition (Stephens and Graham 2010; Bradbury 2016). The Third Global Forum reveals the works in a formal education platform around the world to build its content to support circular economy and system thinking that are required for the transformation (Gitsham et al. 2013). Business schools and corporations/industry alike are advocated as crucial to the development of this new economic model and facilitate education for sustainability, as well as the role of governments related to capacity building and institutional development. Therefore, it is suggested that the design of management programmes through the education framework, particularly in MBA's curricula, could support the shift to a circular economy by allowing students to have first-hand experience, see and learn the inherent value in the products and materials they use from several perspectives. According to Benn and Dunphy (2009), for example, based on the experience of the Dominican University of California, developing green MBA could be considered as a revolutionary approach in bridging the business schools into a more interdisciplinary and sustainable education system. Steiner and Posch (2006) further mention that the case studies which are provided in academics not only proffer the solutions but also help the students outline and define the problems in a complex system. Furthermore, a research conducted by Sky

Future Leader (2011) revealed that 72% of employees credit their business schools in encouraging them to take a long-term view on sustainability and 70% of participants agree that sustainability can create new opportunities for businesses. They can identify how the company works and their impact on the environment in the current and foreseeable future. As it is important that future business leaders are economically literate and able to understand different approaches to the emerging economy realities, universities could be considered as primary sites to enable networking and also a well-paced and flexible education experience (a combination of practical experience and learning). Green economy and the integration of sustainable development into education curricula will thus play a vital role in businesses and policy reforms (Green alliance 2012).

In Indonesia, MBA programmes – equivalent to Master of Management – have an important role in creating future professionals, particularly in regard to the corporate ladder. A significant increase of Master of Management graduates can be seen from 8191 graduates in 2005 to 10,384 in 2007. By 2009, there were 59,021 graduates, of which 15,583 graduated from an A accredited business schools, about 10% of the total university graduates already working in 2014. It should be noted that other than the core content, most Indonesian universities offering Master of Management, particularly in Jakarta, provide for the specialisation to meet the needs for managerial positions. Considering Jakarta's role as a political and economic centre – more than 70% of Indonesian economy is concentrated in Jakarta – it serves as a suitable location for business schools and potential future businessmen. Indeed, out of the 30 universities located in Jakarta, 6 are leading business schools, which are ranked among the best in Indonesia, offering various qualification and study programmes.

Whilst business schools can play a pivotal role in the transformation by promoting the adoption of circular economy initiatives, they face barriers in terms of implementing them in their own operations. There is also a challenge in relation to getting around the pressure of how to ensure university framing of sustainability. To some degree, they explore and engage with sustainability by developing some priorities, based on their own conceptualisation. Typically, focus is placed on ISO certification, green design and other initiatives to address their target to be a sustainable university. Since the terms and language around sustainability and circular economy, however, are complex and endlessly changing, it could lead to different context and adoption, indicating different directions and interests. Therefore, the way the different universities undertake sustainability strategies varies from one another.

Whilst circular economy is rapidly getting popularity around the world, and some countries such as the UK, Australia, China, and the Netherlands have been on the frontline, leading in circular economy best practice, many Indonesian universities lag behind. This leads them to fail presenting real opportunities to design circular sustainability thinking at the heart of the education system. Moreover, the call to implement circular economy models requires breakdown the traditional walls of internal university functions as well as the external sectors such as corporations, suppliers and even other universities. A fact base for Indonesia contemplating the transition to a more circular approach is the lack of recognition by most Indonesian corporations regarding the company's responsibility in social and environmental perspectives as prescribed, for example, by the law of PP No 47/2012 regarding the corporate social responsibility and also UU No 40/2007 regarding the responsibility of limited company. As such, business schools are often accused of moving slowly towards the circular transition.

Within this context, there is a need to understand the best practices with circular university models. The Netherlands is one of those pioneering circularity concepts. Combined with its close relation with Indonesia for hundreds of years, it could be seen as an important source to provide multiple value and opportunities that are likely to benefit Indonesia in future gazing of management programme in circular sustainability. The Netherlands' approaches to turn the concept into reality and tackle the challenges of circular entrenchment in its higher education system could proffer Indonesian universities to focus on moving into circular transition and also tap the opportunity in creating future business leaders from the classrooms.

This paper aims to set future direction of sustainability pathway(s) to incorporate circular economy within Indonesian high education systems and to "reinvent" the university role whilst shaping future leaders. For this purpose, this study intended to develop the definition of circular sustainability high education and introduce a general framework along with its indicators and strategies that can be used to reach such identification. In this regard, this study combines a qualitative and quantitative method by processing the questionnaires using PROMETHEE analysis. The studied cases are two business schools in the Netherlands, Rotterdam School of Management (RSM) and Nyenrode, as well as three leading Indonesian business school/universities, BINUS Business School, University of Indonesia, and Prasetiya Mulya Business School. The main findings, the results of the comparison between Dutch and Indonesian practices as well as the main challenges are presented, and conclusions are drawn. The methods used to collect data are document analysis and in-depth interviews with ca. 40 various stakeholders from the studied universities.

8.2 The Concept of Sustainable University

In a sustainable university, the commitment among all stakeholders strongly influences the institutions. According to Lindsay (2003), a sustainable campus community should actively engage the knowledge of the university community in order to address the ecological and social challenges that we face now and in the future. However, this concept should be translated into a more quantitative index which is measurable and attainable. Velazquez et al. (2006) defined a sustainable university as a higher educational institution, a whole or a part, that addresses, involves and promotes, on a regional or a global level, the minimisation of negative environmental, economic, societal and health effects generated in the use of their resources in order to fulfil its functions of teaching, research, outreach and partnership and stewardship. This way they can help society make the transition to a more sustainable lifestyle. With this drawback, setting an agenda for a sustainability model is not the end of the process. This framework model should be merged within faculties, students and also staffs into a systematic transformation (Hooi et al. 2011).

However, without the assessment indicators on how to monitor the results, the process towards sustainability could be at stake. Geng et al. (2012) also mentioned the necessity to establish the social indicator along with the environmental and economic indicators. The indicators should be explicit and understood by all university levels, and institutionalisation of this idea into the system's culture and its daily operation should be done (Lozano 2006). Alshuwaikhat and Abubakar (2008) advocated that campus operation, research, teaching and efforts to conserve natural resources are the foundations for monitoring and meeting the function. It remains to be seen whether the implications of abstracting the idea of sustainable university by university managers apply towards environmental aspects over economic and social benefits (Wright 2010). To date, based on recent studies, the circular economy could qualify as the next major circular sustainable university model, to be regarded as a "green university".

Adopting circular economy principles does not entail only environmental and social benefits but can also generate economic benefits. The concept in many ways emphasises the transformation to create more value from resources, lowering environmental costs, increasing consumer convenience and securing supplies (Preston 2012). Several relevant experiences have already been recorded: Roy et al. (2008), for example, revealed that the environmental impacts of distance learning in higher education (HE) courses involve 87% less energy and 85% lower CO2 emissions compared to the full-time campus-based courses. Similarly, Karavezyris (n.d) captured that waste management is linked to the more general system goals of resource efficiency and climate protection in a manifold way. Mason et al. (2003) also yielded the waste operation, recycling and educational campaign as an indicative for a zero waste model campus. Pursuing this opportunity, Davis et al. (2009) also rested the concepts on behaviour attitudes, recycling and waste minimisation along with the energy and water efficiency. As part of that effort, Nejati and Nejati (2013) further embedded sustainability practices into teaching, research, community outreach, waste and energy management and land use and planning and scrutinise the definition of a sustainable university as a university that not only to seek academic excellence but also try to embed human values into the fabric of people's lives. Finally, from the case of the National Autonomous University of Mexico, it is expected that when well integrated, new implemented technologies and facilities could address 7.5% less energy consumption and 11.3% fewer GHG emissions by 2020 (Escobedo et al. 2014).

8.3 Green University

8.3.1 A New Definition

Building upon the referred journals and other literature reviews depicting the circular economy practices and sustainability in higher education, this study pinpoints the salient points to help identify the ongoing transformation process to circular principles. As above-mentioned, there is no unified platform of university strategy regarding how to embed the entire value of circularity and sustainability in all of the processes, as these terms are likely to be interpreted in a wide range of adoption and models. To help identifying the intrinsic value and weighty points, this study takes into account the essential characteristics that should be considered as part of a circular sustainability model. It opines that those cannot be served as a trend, but rather as setting the right ideas which help to construct the framework of green university.

The shift to a green university requires to build a more suitable integrated definition that either replaces the existing sustainable university terms or seizes new opportunities to explain what a circular sustainable university is. This study extracts from the essential characteristics and then zooms in to a new definition of green university. A green university is defined in this paper as "a university with integrated management approach aimed at managing continuous changes into sustainability and resource efficiency and be able to deploy academic excellence". This is believed to be restorative of the university system, as well as impactful for society. It is opined that this definition could remedy the limitation of the current term of sustainable university and a just "regular" green university. A regular green university refers to those which seems to focus on sustainability solely (e.g. with focus on ISO certification, green design), not yet fully involved with circular economy adoption. It is the standpoint of this study that in order to have a successful green university operation, an integration of the entire aspects of management approaches, continuous changes, sustainability and resource efficiency, as well as drawing a chance to entrench academic excellence within, is called for.

As such, the function of education managers and their commitment as a management approach are considered essential to the integration and outlining of the vision of this principle at the entire level of university operations. It is about the leadership and how to shift the paradigm towards circular sustainability attitude and place it as the heart of institutions within their instruments, i.e. vision, mission and strategic planning. Furthermore, setting up the transition should not be seen as a long-term or short-term goal, but as a desire ongoing process which will be always pursued in their prolonged period of operation. It is opined that this philosophy of never-ending process should be often made through incremental steps. It can help a university to measure and to communicate its achievements periodically. Resource efficiency is also a matter of green university to make this life cycle thinking as an eminence in their operation. It delineates the connection between the inside system of a university and the emerging world. The idea is having a continuous cycle through the university process and using the unwanted products/process for new activities, also in addition to making people aware of their consumption patterns, thus minimising their ecological footprint to expedite the transition to a more circular world. Ultimately, this paper argues that academic achievements can play the role of catalyst to have simultaneous change. Linking circularity to academic achievements and exposing business students (as future business leaders) to these concepts in practice might affect these future business leaders better prepare them to future challenges. Therefore, green university is seen as manifesting and designing the circularity and sustainability as cornerstones in the education from teaching to research and operations. As a result, a green university will generate students with truly relevant skills to face today's challenges and have the mandate to perform an academic excellence.

8.3.2 Green University Criteria

To recognise the important aspects for developing a framework for green university, this study builds up on the existing research conducted by Velazquez et al. (2006) and comes up with the potential priority factors to benefit its operationalisation.

Management, education and learning, research and valorisation are among the criteria which are opined to have unique value in a manner to sustain circular practices in university systems and be able to educate future business leaders to act in the long-term interest of society and business. These criteria refer to a condition by which the achievement of green university can be assessed over time. This concept also introduces a set of measurements consisting of indicators and strategies, as presented in Table 8.1, to bridge a university capturing the images and infusing them into the entire university operation. The term "indicator" refers to the indication to meet the criteria; "strategy" means a way to attain the measure.

As shown in Table 8.1, this scheme also offers a common heading and guiding principles to translating the green university criteria into a set of measurements to quantifying and assessing the progress. In addition, this concept would allow changing the existing environment into new practices within green university umbrella by encouraging people within the university system, i.e., the education managers, management staffs, researchers, lectures, students and also other stakeholders, to work together and create innovative approaches in establishing circular sustainability.

8.4 The Dutch Practices

Although only two Dutch universities were approached, the analysis gives some insight into what might be the base of the Netherlands current practice and how circular sustainability could be perceived in the level of existing systems and the

Table 8.1 Criteria of green university

Criteria	Indicators	Strategies
Management	Indicating commitment to strategising sustainability	Creating a sustainability culture
		Administering specific function to managing and monitoring the sustainability integration in the university system
	Featuring campus development	Committing to green campus activity and resource efficiency (result-oriented PSS model)
		Striving to have a sustainable process in the value chain of campus operation such as treating the waste according to circularity, energy efficiency, calculating CO2 emission, green building and also organising sustainability events, etc.
		Maintaining green procurement process
	Framing sustainability competences	Having environmental management standards such as environmental management standards/ ISO, utilising framework for business schools such as UN PRME, TBL (Three Bottom Line) and other awarding schemes and recognition
Education and learning	Integrating into the curriculum in formal and informal education programmes	Incorporating the circularity concept into the curriculum and completed with complex modules supporting interdisciplinary and cross-boundary thinking
		Integrating students' soft skills and general management know-how into the course modules
		Develop piloting project and supporting joint programmes
		Encouraging pro-environmental attitudes and behaviour, including the awareness campaign
		Encouraging to develop sustainability specialisation for graduates within circular sustainability concept
	Developing sustainability agent/ambassador	Providing sufficient space encompassing social and personal competencies
Research	Strategising sustainability research	Develop a sustainability research strategies
		Give an account to foster transdisciplinary research and cross-boundary thinking
Valorisation	Managing communicating sustainability	Develop communication strategy for circular sustainability
		Building strong corporate connections
	Community outreach	Social engagement focuses on community involvement

Table 8.2 Important aspects in Dutch education practices

	The differences	
Similar practices	RSM	Nyenrode
Indicates different goals to reflecting circular sustainability but having common to put forward innovation and critical thinking at the heart of this process	Emphasising to adopt green procurement	Having a leverage in sustainable marketing
Having a clear vision and mission and embodying in sustainability statements	Assigning sustainability officer and introducing sustainability ambassador	
Transforming within institutional structures and allow the academic functions to translate the ideas in their own fields	Developing cradle to cradle for innovation and quality	
Collaborating and grounding sustainability and circularity in the wide area of a university system, including within all campus operation		
Develop ongoing awareness campaign and encouraging a wide range of participation		
Critically revise the academic curricula to compile with up-to date situation		
Focusing researches into transdisciplinary approach and cross-boundary thinking		
Making valorisation as a focal point in circular sustainability to make research come to live		

tendency of their internal instruments. To better understand the key inputs, Table 8.2 presents the highlights of the actions to enhance circularity, of both RSM and Nyenrode in their various resource areas. The information then was framed and assessed in a way that captured their comparative practices based on the scoping criteria which was figured in Table 8.1.

It can be inferred from Table 8.2 that the role of university managers is essential for education institutions to build a direction and translate it into actions towards the transition. It is found that both RSM and Nyenrode have already developed and integrated approaches to enact circular economy, advocating it as a central concept, at the heart of the education system. It seems rather difficult, however, to compare benchmarks of each other in the decision-making process. It is growingly recognised that procurement and valorisation are increasingly powered by the circular moves in their system. Thus, it can be underlined that the idea of designing circularity and sustainability is seen more as a philosophy and way of thinking rather than being part of environmental movement solely and is embedded in the entire process within the university activities. The Dutch practice is compared with the Indonesian one in Table 8.3. The idea is to provide insight in which aspects the Dutch experiences can help Indonesian universities to ground the green university concepts within an education institution.

Table 8.3 A comparison between Indonesian and Dutch practices

Viewpoints	Similarities	Differences		
		Prasetiya Mulya Business School	Binus University	University of Indonesia
Vision, mission and sustainability strategies	Differences in translating sustainability into individual management disciplines due to variety of terms in use. Thus, it is found they indicate different goals for sustainability. In addition, it is inferred that the understanding is closer to linear (3P; People, Planet, Profit) rather than circular			
	Does not stated clearly in vision and mission, but it is brought around all university activities. Some of them indicate sustainability in programme objective. However, the implementation varies at different levels			
	The university and business schools indicate good efforts to strive circular sustainability; however, it differs in how to see and reflect the concept into its system	Setting up quality assurance division	Setting building management division and develop central research centre	UI Green Metric Ranking of World Universities
Management, including campus operation	The business schools identified are aware of challenges and barriers to the integration of sustainability			
	The commitment to sustainability, shown into infrastructure, not yet emphasise on CO_2 emission reduction, waste and water treatment. However, indicating the implementation of resource-saving programme through awareness campaign, such as for water and electricity. Not yet treating the waste	New building (for the staffs and classes) in Edu town	Green building (for the staffs and classes) in Alam Sutera, along with the water purifying system	Library (crystal of knowledge) at Depok with a sustainable concept

(continued)

Table 8.3 (continued)

Viewpoints	Similarities	Differences		
		Prasetiya Mulya Business School	Binus University	University of Indonesia
	Does not have specific function to maintain and coordinate sustainability process within institutions, though some of them developed quality assurance division			
	Indication effort to get sustainability competencies beside the accreditation of DIKTI and BAN PT		ISO 9001	
	Stimulus to foster change such as offering grants to qualified staff/students are not yet very common			
	Procurement system is still based on the price, not yet adopting sustainable procurement			
Education and learning	Embedding circularity and sustainability in curricula, though the degree of implementation varies			
	Business schools tend to revise the syllabus of individual modules, few of them trying to embed sustainability across the entire curriculum			
	Have not developed sustainability representative among students and staff to encourage and step up the understanding among students and staffs			
	Sustainability which integrated in the educational programme indicates different gap between postgraduate and undergraduate levels. It is inferred that business schools have the tendency to put more effort in integrating sustainability into postgraduate level.			

(continued)

Table 8.3 (continued)

Viewpoints	Similarities	Differences		
		Prasetiya Mulya Business School	Binus University	University of Indonesia
	The most popular methods to deliver sustainability are guest speakers, field-based learning experience, case studies and local and global competitions. The content of sustainability-related teaching is centred on topics such as ethics, CSR/ community development and sustainability			
Research	Sustainability is seen as the main research focus but not yet elaborated into a more specific sustainability approach			
	Having established research centres and research groups. The themes mostly about the theoretical themes and problem-solving-driven (the requirement of industries and other interest parties). The transdisciplinary research is found, but it has not strongly developed for the entire course programmes			
Valorisation	The concept of valorisation has not yet been familiar. However, in the implementation, some part of this concept has been captured in students' projects/ joint research programmes, and also the communicating sustainability to stakeholders, including industry			

(continued)

Table 8.3 (continued)

Viewpoints	Similarities	Differences	
		Rotterdam School of Management	Nyenrode
Vision, mission and sustainability strategies	Business schools and university indicates different goals for circular sustainability, nevertheless, they stress on the innovation and critical thinking.		
	Stating clearly in vision, mission and sustainability statements		
	The university and business schools indicate good efforts to strive circular sustainability; however, it differs in reflecting the concept in its system	Embedding in the university system, and also setting sustainability goals	Embedding in the university system
Management, including campus operation	Engaging with faculty and providing an account of where it stands on circular sustainability without seeming to interfere with academic freedom		
	Business schools are committed to the idea of circularity and sustainability. Adopting in wide areas, such as energy efficiency, waste and water treatment. The commitment has been spread through awareness campaign		
	Sustainability as a process, considered to be more effective when appointing certain person to administer this process	Sustainability manager	
	Business schools appear to realise the full spectrum of opportunities arising from participation, particularly of staff and students		
	Procurement becoming an important factor in campus sustainability		

(continued)

Table 8.3 (continued)

Viewpoints	Similarities	Differences	
		Rotterdam School of Management	Nyenrode
Education and learning	Striving to embed circular sustainability across the entire curriculum		
	Business schools tend to develop new programme or courses to address sustainability or critically revise the syllabus of individual modules		Lecture initiative to incorporating circularity and green marketing
	Attempting the ambassadors and sustainability events among students and staffs to keep them up in engaging sustainability	Sustainability ambassador Sustainability officer	
Research	Transdisciplinary research becomes research focus and elaborates them into sustainability approach		
	Having established research centres and research groups to carry out research dedicated to sustainability and to enhance the overall integration of sustainability in their organisation	Developing centre of research which is giving more concern on cradle-to-cradle aspect	
Valorisation	Valorisation is inferred to be a focal point of circular sustainability. The idea is to make research come to live		

8.5 The Indonesian Practices

The Indonesian interviews reveal a similar result to the one in The Netherlands. It shows that the way Indonesian universities perceive the idea of green university and how they implement it depend on their own systems. It strengthens the idea that basically there is no definite means/prescription to regain and reassert sustainability and circularity in the university system in general. The findings further show that the Indonesian participants are focused more on the narrow perception of the concept of green university. Most of the participants related the green concept with leadership (94%), infrastructure (91%), curricula (89%), energy efficiency (80%), student competencies and attitudes (86%) and also strategy (86%). However, fewer participants (37%) indicated that green concept should relate to procurement process and closed loop system (43%). This result is not surprising since the idea of circular economy has not gained wide recognition in Indonesian universities. Nevertheless, this result has also indicated a readiness to undertake the green university as part of the system to some degree. However, such a commitment to reflect their consciousness seemed to be doubly indeterminate, particularly in the sense that they have to be firmly embodied by the principle. The main findings of the Indonesian practices are presented in Table 8.3. It provides an insight into how the concept of green university is conceived and put to practices in the five cases analysed and compared in this paper based on the selected criteria.

As presented in Table 8.3, in comparison to the Dutch practices, Indonesian universities have already nurtured and seemed to be in tune with this concept, but not focused on pursuing circularity. It is lagging behind the Netherlands, particularly in the aspects of valorisation and the scope of research. For example, despite the review showing that universities/business schools seem to give some recognition to sustainability in their research, it is revealed that there is not much information gathered related to sustainability research strategy. In addition, those researches can be partly characterised as integrative approach and mainly problem-driven. In regard to valorisation, it is revealed that Indonesian universities are not yet at the stage where they are fully able to convey the provision of its operation into the companies'/stakeholders' needs.

Additionally, it appears that there is no consensus on the exact way of implementing circular sustainability. It is up to the university managers to define the direction and shape the implementation process, which is assumed to be the main reason which causes the difference approach in the way of energising this concept within their institutions. The vision of Dutch institutions reveals that linear sustainability – which was being used to define a sustainable university– is no longer tenable. Rather, circular economy (or often called as cradle to cradle) and its proponents are seen to be a different way of thinking, not an environment movement solely. Circularity and sustainability have been used as a measure to envision and articulate the future, which are embraced in a wide area of a university system, ranging from curricula, research, procurement and campus operation to valorisation. This aspect seems to be missing in the case of the Indonesian university; thus, it can be used as a starting point for university leaders to play an important role in coordinating and changing institutions from top-down to bottom-up or even double envelopment to implement circular sustainability at all universities and faculties/school levels.

8.6 The Challenges

The green university has a chance of succeeding if the entire level of university managers and other stakeholders all work together. However, many barriers and challenges are yet to be tackled. With regard to the barriers in Indonesia, three main classifications are identified, (1) lack of direction, (2) lack of motivation and (3) lack of abilities of people within the institutions, all of which could hinder the green

Table 8.4 Barriers and suggested solutions, Indonesian perspective

Barriers	Proposed solutions	
Lack of direction	The uncertainties surrounding the translation of sustainability definition and its goals	Setting goals in circular sustainability for the system which accentuate the innovations and critical thinking
	Have not yet clearly stated sustainability as vision and mission, though at certain level, it has been implemented	Stating clearly in vision, mission and sustainability statements
	The commitment to sustainability is only partly shown	Adopting awarding schemes and recognition in line with the university goals
	Have not yet administered specific function	Conducting awareness campaign addressed particularly to students and staff in resource efficiency
		Embedding the circular sustainability into the whole university system, not only in infrastructure, research and education
		Administering sustainability manager
Lack of motivation	Limited financial and infrastructures supports	Giving financial support and building infrastructure
	Facing limited control over (particularly) customised educational programme (executive class)	Giving more responsibility to develop new programme which embed circular sustainability across the entire curriculum and revise the existing syllabus
	Partnership with clients who might not appreciate the sustainability	Identifying the same path in circular sustainability among postgraduate and undergraduate level
	The implementation of resource-saving programme has not yet covered at whole system	Showing commitment to sustainability and circularity to any related stakeholders involved
		Adopting in wide areas, such as energy efficiency and waste. In the beginning, conducting study to measure the resource use and identify the saving programme would be useful
Lack of abilities of people within the institutions	Staff reluctance to integrate sustainability into their teaching, such as lack confidence, knowledge, time, etc.	Providing stimulus to foster change and encourage them to start the initiatives, such as through sustainability ambassador or sustainability officer of the month
	Mismatch in addressing sustainability within their courses	The discussion on circular sustainability should be transdisciplinary, across the programme and organisation

(continued)

Table 8.4 (continued)

Barriers	Proposed solutions	
	The inconsistency of individual culture and attitude towards sustainability	Creating a sustainability culture
	The recent initiatives to foster change among students and staffs have not yet indicated strong effect	
	The transdisciplinary research has not strongly developed	
	Procurement has not yet considered as important factor	Adopting sustainable procurement
	Valorisation has not yet become a vocal point	Embedding the circular sustainability into the whole university system including valorisation

university concept. Table 8.4 presents the main identified problems as well as possible solutions as a result of mirroring process from the Dutch best practices to re-grounding the idea of circularity and sustainability.

In Table 8.4, an attempt is made to identify some potential solutions to the current barriers presented to the green university implementation in the Indonesian universities. Many of the suggestions are in line with examples found in the literature and obtained from interviewees of the Dutch universities. In the following and last section, the authors draw some conclusions from this research.

8.7 Conclusions

Universities can play an important role in shaping future business leaders to become active in the path towards sustainability by acting as a green university. However, in order to embed the concepts of circularity and sustainability, there is a need to expand the concept of green university to include broader aspects and be approached as a continuous process rather than seen as a short-/long-term goal.

It is clear that Indonesian universities are far behind the Dutch ones in terms of circular sustainability (green university) practices. The Indonesian universities are still focused more on the narrow approach, whereas in the Dutch case, there is a holistic philosophical approach referring to sustainability not as part of an environmental movement but rather as a way of life. There are many reasons and circumstances involved, though this study was limited to the managerial approach to grasp the concept of green university and how the university leaders decide to include it through the green university criteria.

To institutionalise the green university principles, it was found that that university leaders need to ensure the continuity of a clear institutional vision which needs to be instrumented with specific policy strategies in a holistic approach. This can be seen as a first step towards implementing circular practices at all levels.

Although this study was centred on the circularity and sustainability practices at only three leading business schools in Jakarta, it is able to provide significant empirical information on how the Indonesian educational systems at university level perceive and practise those concepts within their system. It also offers a new definition of green university, as well as suggests a general framework. This suggested framework can be used in further investigation and follow-up research, potentially providing a guidance towards the Indonesian university transition.

References

Alshuwaikhat H, Abubakar I (2008) An integrated approach to achieving campus sustainability: assessment of the current campus environmental management practices. J Clean Prod 16(16):1777–1785. https://doi.org/10.1016/j.jclepro.2007.12.002

Benn S, Dunphy D (2009) Action research as an approach to integrating sustainability into MBA programs. J Manag Educ 33(3):276–295

Bradbury H (2016) Sustaining inner and outer worlds: a whole-systems approach to developing sustainable business practices in management. J Manag Educ 27(2):172–187. https://doi.org/10.1177/1052562903251414

Davis G, O'Callaghan F, Knox K (2009) Sustainable attitudes and behaviour amongst a sample of non-academic staff: a case study from the information services department, Griffith University, Brisbane. Int J Sustain High Educ 10(2):136–151. https://doi.org/10.1108/14676370910945945

Escobedo A, Briceno S, Juarez H, Castillo D, Imaz M, Sheinbun C (2014) Energy consumption and GHG emission scenarios of a university campus in Mexico. Energy Sustain Dev 18:49–57

Geng Y, Fu J, Sarkis J, Xue B (2012) Towards a national circular economy indicator system in China: an evaluation and critical analysis. J Clean Prod 23(1):216–224

Gitsham M, Culpin V, Bond D, Ashridge Business School (2013) Building leadership capability for a rapidly changing world: what every business leader now needs to know about developing themselves and their people in a changed world. Ashridge Business School, Berkhamsted

Green Alliance (2012) Green Economy: a UK success story. https://www.green-alliance.org.uk/resources/Green%20economy%20a%20UK%20success%20story.pdf. Accessed Oct 2015

Hooi KK, Hassan P, Jami NA (2011) Sustainable education: an assessment of carbon footprint at UCSI university and proposed green campus initiative framework. Paper presented at the 3rd international conference on information and financial engineering. http://www.ipedr.com/vol12/62-C140.pdf. Accessed Aug 2015

Karavezyris V (n.d.) Circular Economy in Germany: achievements and future challenges. http://www.iswa.org/uploads/tx_iswaknowledgebase/Karavezyris.pdf. Accessed Aug 2015

Linsday C (2003) Assessing sustainability on Canadian university campuses: development of a campus sustainability assessment framework. http://neumann.hec.ca/humaniterre/campus_durable/campus_memoire.pdf. Accessed Aug 2015

Lozano R (2006) Incorporation and institutionalization of SD into universities: breaking through barriers to change. J Clean Prod 14(9):787–796. https://doi.org/10.1016/j.jclepro.2005.12.010

Mason IG, Brooking AK, Oberender A, Harford JM, Horsley PG (2003) Implementation of a zero waste program at a university campus. J Resour Conserv Recycl 38(4):257–269. https://doi.org/10.1016/S0921-3449(02)00147-7

Nejati M, Nejati M (2013) Assessment of sustainable university factors from the perspective of university students. J Clean Prod 48(3):101–107. https://doi.org/10.1016/j.jclepro.2012.09.006
Preston F (2012) A global redesign? Shaping the circular economy. https://www.chathamhouse.org/sites/files/chathamhouse/public/Research/Energy%2C%20Environment%20and%20Development/bp0312_preston.pdf. Accessed July 2015
Roy R, Potter S, Yarrow K (2008) Designing low carbon higher education systems: environmental impacts of campus and distance learning systems. Int J Sustain High Educ 9(2):116–130
Steiner G, Posch A (2006) Higher education for sustainability by means of transdisciplinary case studies: an innovative approach for solving complex, real-world problems. J Clean Prod 14(9–11):877–890
Stephens J, Graham A (2010) Toward an empirical research agenda for sustainability in higher education: exploring the transition management framework. J Clean Prod 18(7):611–618. https://doi.org/10.1016/j.jclepro.2009.07.009
Velazquez L, Munguia N, Platt A, Taddei J (2006) Sustainable university: what can be the matter? J Clean Prod 14(9–11):810–819
Wright T (2010) University presidents' conceptualizations of sustainability in higher education. Int J Sustain High Educ 11(1):61–73. https://doi.org/10.1108/14676371011010057

Chapter 9
Share, Optimise, Closed-Loop for Food Waste (SOL4FoodWaste): The Case of Walmart-Mexico

John Rincón-Moreno, María-Laura Franco-García, Jorge Carlos Carpio-Aguilar, and Mauricio Hernández-Sarabia

Abstract The food waste in landfill decomposes into contaminated run-off (leachate) and methane (CH_4), which is considered a relevant greenhouse gas. This causes environmental liabilities, energy losses and problems in the food system. Currently, organic waste volumes are increasing dramatically converting this into a serious concern in both developed and developing countries. Zero waste to landfill (ZWTL) is one of the most promising concepts for solving organic waste problems. ZWTL when integrated into business processes can lead to innovative ways to identify, prevent and reduce waste. In that sense, the circular economy (CE) has also been considered regularly as an approach to the more appropriate waste management as it considers the business strategy part of the zero waste system. This circularity would increase productivity throughout the food value chain. In that manner, retail stores are proven to be a major market-driven force in the food system. Hence, one retail store located in the Metropolitan Area of Mexico City, part of Walmart-Mexico (Walmex), was selected to showcase a suitable strategy to tackle the food waste issue. Thus, this research aimed to explore how the organic waste management can be improved by combining CE business model and a ZWTL strategy. The findings of the combined framework (SOL4FoodWaste) showed that most of the food considered as waste can be recovered through different stages. Even further, 40% of the

J. Rincón-Moreno (✉)
University of Twente, alumni, Enschede, The Netherlands

M.-L. Franco-García
University of Twente, Enschede, The Netherlands
e-mail: ml.lfrancogarcia@utwente.nl

J. C. Carpio-Aguilar
Walmart Mexico, CDMX, Mexico
e-mail: jorge.carpio@walmart.com

M. Hernández-Sarabia
Faculty of Engineering, Universidad de Ibagué, Ibagué, Colombia
e-mail: mauricio.hernandez@unibague.edu.co

M.-L. Franco-García et al. (eds.), *Towards Zero Waste*, Greening of Industry Networks Studies 6, https://doi.org/10.1007/978-3-319-92931-6_9

food waste management costs can be saved through three business actions associated to those recovery stages. The proposed SOL4FoodWaste framework intends to collect all the sustainable concepts that might potentially be implemented or resembled in other industries with similar food waste challenges.

Keywords Organic waste management · Zero waste to landfill · Food waste hierarchy · Circular economy · ReSOLVE framework · SOL4FoodWaste

9.1 Introduction

At different stages (from harvest to consumption), the food is thrown away even if it is still appropriate for human intake. It is estimated that up to 30% of food produced for human consumption in the world is wasted throughout the food supply chain (FAO 2011). This fact is having significant impact on the current food system posing many uncertainties on environmental liabilities as in many countries organic waste is buried in landfill locations. The food landfilled decomposes into contaminated run-off (leachate) and methane (CH_4) greenhouse gas. Methane is 21 times more harmful than carbon dioxide (Forster et al. 2007) in terms of greenhouse effects. Thereby, CH_4 is linked to global warming but is also related to energy losses due to its heating and cooking properties (Moumen et al. 2016). Therefore, solutions on food waste burdens are essential to improve food security and the efficient use of resources as well. Ultimately, the food waste reduction and/or food waste elimination can enhance the entire food system (Koester 2015).

At the moment, countries are constantly dealing with the food waste issue; so far, the developed nations are regarded as the frontrunners, while the developing ones are still struggling with their waste management systems. As an example of the current situation in developing countries, it has been reported that about 96% of the municipal solid waste (MSW) in Mexico is still disposed of in landfills. Fifty percent of the total of the disposed MSW is attributed to organic components. It is worth mentioning that irregular dumping areas can be used to 'dispose' the waste. Likewise, recycling is not yet embedded in the Mexican culture, and the consumer awareness qualifies low due to lack of proper orientation and public policies at this regard (SEMARNAT 2015). Thus, Mexico's current waste management system is considerably lagging in recycling and energy recovery and offers little regulatory incentive to avoid food being disposed of.

Even so, there are initiatives driven by the private sector that avoid waste disposal. Thereby, this research was focused in one of those practises done in the private sector, more exactly within Walmart-Mexico (Walmex). At present, Walmex's objective is to avoid the generation of food waste whenever possible by seeking out a zero waste to landfill (ZWTL) strategy. Currently, the waste that cannot be recycled, reused or composted is disposed of in authorised site for either urban solid waste or hazardous waste. The waste landfilled still corresponds to 27% of the total waste produced (Walmart 2015), though this company has a sustainable waste management policy that stimulates the waste management efficiency in the retail stores with a special focus on food waste. Even further, it has multiple organisational areas that constantly review the regulatory frameworks in order to comply with permits and authorisations required to operate its sustainable waste management policy. Thus, an enhanced approach on waste management regarding food waste is needed to increase the up-to-date efficiency level. The reduction on the food waste burden is not only seen at the company as a sole environmental sustainable strategy but rather is expected to contribute in a sustainable business through a circular use of resources. Hence, it would generate economic benefits to the business in the midterm, fostering corporate sustainability, environmental awareness and society welfare.

A high potential in preventing food waste by managing resources at lower cost is seen as a promising advance towards a sustainable world (MacArthur 2013). By doing so, the organic waste might return to the natural systems via energy recovery or can be used again with slight loss of quality (Pitt and Heinemeyer 2015). Solutions on food waste such as donation, utilisation of by-products, livestock feeding, composting or bioenergy generation are put in consideration when dealing with food waste. These approaches not only offer social and environmental benefits but also economic advantages (Jurgilevich et al.2016). In this regard, circular economy (CE) is a new business model aiming to close the loop of resources (productive process spill overs and post consumption materials) by keeping their value as much as possible.

While the basic concept of the CE has been available for many years, the benefits of this business model are often misaligned with the core business strategies (Planing 2015). In fact, there are not sufficient case studies that currently reflect the necessity of going circular in business and even fewer examples of companies reporting food waste reductions aligned with their core business strategy. Hence, the reduction on food issues towards a CE model must entail a financial-feasible strategy where no organic resources become landfilled.

For this particular study, the focus is the circular management of food waste. Thus, this study aimed to elucidate a framework that can improve the organic waste management in Walmex by combining a circular business model and a ZWTL strategy. In this study, the (CE) principles applied to this type of organic materials were translated to four management streams: (1) reduction, (2) donation, (3) animal feeding and (4) anaerobic digestion. Later on, all of them were incorporated into a circular business model for the case study in hand.

The research is structured as follows: the steps needed to reach a CE business model through a ZWTL strategy was detailed in Sect. 9.2. Then the methodological approach undertaken was described in Sect. 9.3. The findings and discussion as a result of the methodology applied to the case study were explained in Sect. 9.4. Finally, conclusions and recommendations were placed in Sect. 9.5.

9.2 Literature Review

This section details the concept of organic waste management, especially focused in developing countries to provide an overview of the current situation. Then the ZWTL was explained in a food waste context. This could be possible by introducing the food waste hierarchy as an approach of targeting the ZWTL strategy. Finally, the CE principles were coupled into a suitable framework for this case study through the ReSOLVE framework.

9.2.1 Organic Waste Management

Organic waste volumes are increasing dramatically due to economic growth, rise in population and higher standards of living converting this into a serious concern in both developed and developing countries (Kothari et al. 2014). In developing countries, this has been one of the major problems on waste management today due to the lack of funding for waste treatment, insufficient land for disposal and low expertise coupled with weak policies and regulations interfering with the implementation of an appropriate system (Mohee et al. 2015).

Currently, five popular treatment methods are widely used in developing countries to treat organic waste: animal feeding, composting, anaerobic digestion, incineration and landfills. Illegal open dumps and landfills are primary methods due to their widely application for treating food waste (Adhikari et al. 2006). Food being disposed of in landfills represents 90% of use rate for food waste treatment, and the second most common method, composting, ranges from 1% to 6% of use rate. Anaerobic digestion (use rate of under 0.6%) and other treatments, such as livestock feeding or incineration, are rarely used (Adhikari et al. 2009).

In order to implement a different perspective on organic waste management, it is important to categorise the waste composition into different categories such as recyclability, combustibility or biodegradability opportunities so it can be used as a parameter to implement the most suitable waste management strategy. Nevertheless, when choosing the adequate system, some criteria needs to take into consideration several variables for an effective waste management such as the primary service and secondary services/coproducts provided such as food for donation, livestock, composting or anaerobic digestion (Villanueva and Wenzel 2007).

Although, those options seem to be the most feasible solution, it may not be convenient for technical or economic reasons in some cases (Jamasb and Nepal 2010). Overall, in essence, organic waste management must be focused into the principle of 'zero waste' which means no raw materials being disposed of to landfills throughout the food chain (Song et al. 2015).

9.2.2 Organic Waste Management in Mexico

In general, the Mexican organic waste recycling system ranks very poorly compared to similar economies (SEMARNAT 2015). This is due to incomplete legal frameworks and limited funding to improve the food waste systems. Another relevant factor is the little consumers' participation in the collection phase; inappropriate education programmes at this regard seem to be the cause. Thus, collection of recyclables remains very weak across most locations in the country (Thi et al. 2015).

According to SEMARNAT (2015), 96% of the municipal solid waste is sent to landfills; very insignificant portion (3.6%) receives some type of recycling. In addition, open dumps and inadequate landfills are the most common practise. At present, there are 12 active sanitary landfills in Mexico, but it is known that there are numerous undocumented landfills especially in rural areas. In any case many landfills in Mexico are reaching their capacity thresholds, and municipalities are seeking funds to provide new solutions. Despite the government's effort to support these initiatives to upgrade infrastructure, not substantial improvements have been observed.

A high percentage of organic waste (up to 52% of the solid waste) that is disposed in open dumps or quasi-controlled landfills (SEMARNAT 2015) is an indication of the inadequate waste management system. This generates an increasing environmental impact as no treatment is being developed to solve the problem on organic waste management. The environmental problems are associated with methane emissions, bad odours, leachates, etc. (Moumen et al. 2016). In addition to the issues associated with the large physical area required for the landfill facilities, in practice, the landfills' capacity is often surpassed. Although 48% of the municipal solid waste represents different types of waste that can be easily recovered, the focus is on the organic waste. It represents a twofold solution: feeding opportunities and energy recovery.

9.2.3 Waste Management at Walmex

Current Walmex's waste management actions are framed under the ZWTL strategy and the local environmental regulations. So far, the waste treatment methods at Walmex are determined according to the existing infrastructure and its potential; residues recovered with economic value are already sent to recycling. Composting is determined according to the available infrastructure at the region where shops are located. The residues that are not recycled, reused or composted are sent to authorised disposal sites for waste urban or hazardous solid waste. In general, 355,944 tonnes of waste are generated in Walmex, from which 73% is recycled and 27% is sent to landfill (Walmart 2015).

9.2.4 Zero Waste to Landfill (ZWTL)

Exhaustion of finite natural resources forces society to consider efficiency on their industrial outputs. Therefore, ZWTL is understood as a philosophy that inspires to reshape resource's life cycles so no junk is being disposed of either in landfills or incinerators (Song et al. 2015). Some scholars might consider that 'zero waste' management is an all-inclusive view of waste and resource treatment from a sustainable standpoint. Essentially, ZWTL aims to resemble the way resources are reused in a natural cycle. It promotes a transformation, where businesses minimise the burden on earth's raw materials and learn to do more with less natural resources (Song et al. 2015).

Certainly, the scope of ZWTL embraces concepts that have been settled for sustainable waste management systems which comprise avoiding, reducing, reusing, recycling and regenerating waste materials (Zaman and Lehmann 2013). However, ZWTL embodies integrated systems in which everything keeps its value. Thus, the material stream becomes circular meaning that the same materials are used again and again until they are exhausted. No materials are wasted or underused in a circular system (Colon and Fawcett 2006; Mason et al. 2003; Murphy and Pincetl 2013).

ZWTL is one of the most promising concepts for solving waste problems. Its concepts depict an ideal environment and resource utilisation. Irrespective of the social, economic and environmental benefits of ZWTL, the concept itself provides gains for an organisation or community if they channel the work in a clear and consistent manner (Song et al. 2015). When ZWTL is integrated into business processes, it can lead to innovative ways to identify, prevent and reduce waste of all kinds; albeit, in order to know whether they reach the ZWTL target, how to evaluate the process becomes very important.

9.2.5 Waste Hierarchy

Nowadays, there is no straightforward path for solving the current waste challenges. Hence, it cannot be expected to have a common framework on handling waste within industries, nations and households. Though, especial attention to managing waste has been placed on legislations, action programmes and setting the targets to embark on this problem. The Directive 2008/98/EC of the European Parliament (2008) presents a framework that aims to diminish the undesirable effects to health and environment of the production and management of waste.

Hitherto, the framework elaborated by the European Parliament stated that waste prevention shall be the main concern of waste management, then, reusing, material recycling and energy recovering shall be the least desired option. The EU directive layouts the waste hierarchy as an instrument on handling waste. However, it is highly ranked as a practice to not support landfilling.

9.2.6 Food Waste Hierarchy

Likewise, in the worldwide waste challenge, the increasingly food waste issue is drawing more attention. Hence, a conceptual framework depicting the waste hierarchy proposed by the European Parliament in the context of food waste is relevant in order to guide the most suitable options under this perspective. As an illustration, authors such as Papargyropoulou et al. (2014) proposed a framework for the management of waste all the way through the food supply chain by adopting the EU waste hierarchy framework. The authors suggested a twofold categorisation to distinguish what can be recovered as food surplus which in this scenario is considered as waste avoidance (e.g. reduction, donation and livestock feeding) and what can be used as food waste for the recovery step (energy recovery, e.g. biodigester). This definition was also studied by Parfitt et al. (2010), whose definition of food waste is related to the reduction in fit-for-human consumption organics at the end of the food supply chain (retailers and consumers), which means that it can no longer be eaten.

As previously mentioned, setting the priorities and ways of management is crucial as very often the main measure is the reuse of surplus food for human consumption to relieve food poverty. This latter is mainly organised through redistribution networks and food banks. Yet, the quality of food surplus might undermine those approaches, and other treatments might be included to process food waste. Those suitable solutions are explained in the following sections.

9.2.6.1 Reduction

The reduction of organic waste is being increasingly seen as a concern by corporations, organisations, governments and so on. This is due to the outrageous amount of food landfilled from farm to fork. For this reason, the target is firstly being set on preventing overproduction and oversupply of food beyond human diet requirements at all the stages of the food value chain. In the retail market, it also entails food surplus prevention (Papargyropoulou et al. 2014). Food waste could be easily reduced through inventory management to minimise surplus while maximising shelf replenishment and in-store shelf life times (Bond et al. 2013).

Some strategies directly affect food losses and food waste reduction, while others are indirectly affected by influencing people's behaviour in a long-term strategy. Hence, it is important to optimise the flow of perishable food at its very primary stage taking into account the expected shelf life of the food to prevent needless losses. This optimisation might bring minimisation of economic deficits, while consumers can be provided with high-quality products (Blackburn and Scudder 2009). Any sort of approach undertaken to address this issue must be particularly practical and cost-effective, so that the implementation could be performed relatively quickly and could achieve short-term gains once it has been stablished (Lipinski et al. 2013).

9.2.6.2 Food Donation

Food donation to those in need presents interesting opportunities since there are food networks that may contribute to this ideology by offering more easily controlled systems avoiding food loses. Therefore, a link is created to associate sustainability with food security and food safety. Though, there are always losses in the food chain no matter how close the loop is, it is also necessary to consider the obstacles when generating solutions for food waste management when donating (Jurgilevich et al. 2016).

9.2.6.3 Animal Feeding

It is based on the principle that the food waste can serve to feed the livestock to foster the agricultural output. In some Asian countries (Japan, south Korea, Taiwan) with high demand for animal feeding, it is promoted by law the use of food waste to feed animals, which accounts for almost half of the total food waste generation (Kim et al. 2011). Albeit, in developing countries it is not a common practise to separate and collect food waste to feed animals since almost all of generated food waste is mixed with municipal solid waste (Thi et al. 2015).

9.2.6.4 Anaerobic Digestion Technology

It refers to the biological decomposition of organic materials in the absence of oxygen to produce a biogas mainly methane and a fraction of carbon dioxide with an effluent of high nutrient value (Ray et al. 2013). Anaerobic digestion has been widely investigated for both its energy production and waste management potential. This is a well-known technology that has been implemented in various countries of the world (Thi et al. 2015). On the other hand, it is acknowledged in developing countries that anaerobic digestion is still hard to implement due to technical barriers, inappropriate operations or inadequate management (Müller 2007).

Food waste is well-fit to be converted into methane by anaerobic digestion, and some studies have been carried out evidencing its effectiveness (Wang et al. 2012). Under anaerobic conditions the food bacteriological content converts the food waste into biogas. Such application is attractive due to its potential for energy (heat and electricity) generation (Chiu and Lo 2016). Compared with other typical food waste treatment such as composting and incineration, anaerobic digestion is more appropriate for food waste treatment because it recovers more energy while it emits less environmental harm in the form of greenhouse gas emissions (Chiu et al. 2016).

The biogas derived from the anaerobic digestion can be used as an alternative energy source to substitute conventional fossil fuel consumption (Calabro 2009). On this manner, it can be used as various types of energy sources, such as heat and electricity produced from biogas combustion (Lin et al. 2013). The heat can be used to uphold the temperature of the anaerobic reactor and to provide heat for nearby areas. If electricity is generated, this can be internally consumed, and the excess of electricity could be delivered to the public grid (Møller et al. 2009).

In order to reach a 'ZWTL' strategy, a solution is not completed until all the factors affecting not only the environment but also the social and economic factors are counted. In consequence, the 'ZWTL' philosophy integrated into the food waste hierarchy is one approach that must be coupled with a way to create a cost-effective business with social benefits. Thereby, the CE principles offer a framework where the ZWTL strategy can be valued towards a circular business model as explained in the following section.

9.2.7 Circular Economy (CE)

CE is an approach that takes the concept of sustainability one step further and rethinks the future on how the current system can be reorganised, where growth is possible within a renewing economy involving materials science, design, engineering and systems design (Pitt and Heinemeyer 2015). In essence, CE is known as an industrial system that is meant to replace the end-of-life concept and the use of renewable energy and targets the elimination of waste through the new output of products, organisations and even business models.

This CE system is focused on few principles. The first principle of CE aims to preserve and improve natural resource management and renewable resource streams over finite stocks. The second principle introduces the optimisation of resource by circulating products, components and materials in use at the end-consumption utility in both technical and biological cycles. Finally, the third principle considers fostering system effectiveness by designing out externalities (MacArthur 2013).

Companies seeking for savings in circular systems need to increase the rate at which their products are collected and consequently reused or recuperated by applying CE principles. As it can be seen, a circular model fosters the need for innovation and growth and decreases the demand for resource consumption. Productivity can be threatened when natural resources become scarce; thus governments and companies have started considering the implementation of CE business models (Murray et al. 2015).

As it has been reported (MacArthur 2013), increasing circularity creates new opportunities for companies that could offer important benefits that would not otherwise be achievable without a CE model. Through different schemes, companies could create strategies to face business challenges by starting up new CE models so that bigger profits will appear, for example, by making use of by-products. As a result, there will be a lower energy and material intensity demand, thus reducing fossil fuel consumption (Murray et al. 2015). Moreover, the lower demand for resources will also slow current rates of natural erosion relieving environmental burdens and decreasing the need for landfill and charges for waste treatment.

According to the Ellen MacArthur Foundation (2013), the full benefits of the CE would avoid up to 340 million tons of waste from landfilling each year around the globe, more than 272 million tonnes would be from the elimination of food waste. This circularity would reduce the pressure on agricultural soils and the preservation of land productivity through less pressure on food production. Given that around one-third of the total food produced for human consumption is now lost or wasted (FAO 2011), using this 'waste' material has considerable potential to slow the demand for land.

CE has been considered regularly as an approach to more appropriate waste management through recycling, reuse or recovery methods. Hence, a complete view at the implementations of alternative solutions over the entire life cycle of any process as well as the interaction between the process, the environment and the economy must be considered (Ghisellini et al. 2016). After all, CE has the potential to implement new patterns and help stakeholders reach increased sustainability. Consequently, it is not only regarded as a waste or energy recovery, but rather it is about improvement of the entire system.

9.2.8 ReSOLVE Framework

According to the Ellen MacArthur Foundation (2013), the three principles of the CE above outlined can be applied into six business actions. Those actions might contribute to businesses and countries to generate circular strategies and growth initiatives. Thus, it represents circular business opportunity under the so-called ReSOLVE framework (Regenerate, Share, Optimise, Loop, Virtualise and Exchange). This conceptual framework displays a major role as it can be used to apply the principles of the CE towards the design, implementation and acceptance of circular business models (Lewandowski 2016; Planing 2015).

The ReSOLVE framework is briefly detailed as follows:

(1) Regenerate – primary use of renewable energy and materials, involving health regeneration of ecosystems such as return of the biological resources to the biosphere.
(2) Share – this means optimisation of resources by sharing them among users and reusing them throughout their lifetime through maintenance, repair or refurbishment.
(3) Optimise – this refers to a performance/efficiency of a product by removing waste in production throughout the supply chain.
(4) Loop – it is meant to keep components and materials in closed loops by prioritising inner loops.
(5) Virtualise – this is related to delivering goods and services virtually.
(6) Exchange – it replaces old materials with advanced non-renewable materials by trying to apply either new technologies or new products and services (MacArthur 2013).

9.3 Methodological Approach

As mentioned in the introduction, the main research question driving this work is how the organic waste management in Walmex can be improved by combining a CE business model and a ZWTL strategy. In order to study the CE business model and ZWTL frameworks, it was decided to use the case study approach. In addition to that, desk research, interviews with semi-structured questionnaires and expert's consultation were combined for data gathering. The analysis of the information was under the combination of those frameworks and the food waste hierarchy.

9.3.1 Case Study

This research aimed to provide a suitable strategy to closing the loop on food waste at Walmex. This was done by integrating the 'ZWTL' philosophy into the ReSOLVE framework proposed by the Ellen MacArthur Foundation (2013), as the conceptual framework method employed towards a circular business model. The showcase in this study was based in one Walmex located in the Metropolitan Area of Mexico City.

9.3.2 Data Collection

In order to receive a complete input of the performance mainly on food safety, food quality and food waste management, key staff members in the store were consulted. The authors conducted semi-structured questionnaires among staff members with positions such as store manager, reception supervisor, assets protector supervisor, IT supervisor, bakery supervisor, fruit and vegetable supervisor, maintenance leader, operations submanager, perishables submanager and canteen manager. In total, 12 respondents were selected due to their influence in the decision-making processes throughout the food value chain at the retail store level.

The questionnaire was applied to key staff enquiring the same questions to all of them but with slightly variations according to their level of involvement on each step of the current organic waste management practises (e.g. managerial decision questions, operational performance). Afterwards, this sort of outcome was coupled with information of daily food waste generated and the type of waste generated. Both characterisation and volumes were gathered through logbooks provided by the staff which served as the inventory baseline of the current practises. This methodology was selected and carried out due to its double character: qualitative and quantitative nature.

9.3.3 Data Analysis

The current organic waste management was analysed against the food waste hierarchy proposed by Papargyropoulou et al. (2014). This was done due to the practicality of using a framework to assess the practises that influenced the efficiency of the store. Given the different steps in the hierarchy before landfilling, the data collected was assessed in terms of food waste avoidance (reduce, feed people in need and feed livestock) followed by waste management (energy recovery, e.g. biodigester) as described in Sect. 9.3. However, the analysis of the information only through the waste hierarchy was limited to the evaluation of the environmental performance. Consequently, in order to integrate the technical, environmental and financial impacts, the information gathered was analysed through the CE principles set by MacArthur (2013) as described in Sect. 9.4.

9.3.4 Circular Business Model Analysis

The findings described in this study are seen as an input towards a circular business model. In that sense, the integrative approach to reach the aimed target was assessed under a multidimensional perspective, i.e. technical, environmental and financial. This integrative approach shed light on the benefits generated that can increase the acceptance of CE models and practises. The integrative perspective was offered by the ReSOLVE framework which provided the conceptual background from which the ZWTL is shifted into a business action, as it was presented in Sect. 9.4.

The two main steps followed during the ReSOLVE framework implementation are here indicated. Firstly, the strategy proposed was confronted between two scenarios carried out through the ReSOLVE framework, being the inner circle of the loop and the priority given because it guarantees a longer and more efficient use of resources towards an integrated circular business model. Then, the technical-environment-financial aspects were the ones that determined the decision-making variable in this study.

9.4 Findings

The interviews were carried out among 12 key responsible persons at the retail store. The enquiries on the subject of organic waste management activities provided evidence on how those activities influence the efficiency of the system. This resulted in an inventory of the current practises implemented at such location. Moreover, in order to keep details of food waste, logbooks were consulted to track the type of food and daily volumes of organic waste disposed of to landfill. Subsequently, the findings were used as baseline to be integrated to the ZWTL strategy with the purpose to improve the current organic waste management system.

9.4.1 Current State of Organic Waste Management at the Walmex's Store

At present, 'first-in, first-out' is the adopted systemic solution in the store which is based on the assumption that all food products arriving on a specific time have the same shelf life span. Even so, the store's policy regarding this issue stated that despite this fact, store shelves have to be well filled with a wide range of fresh products to anticipate customers' expectations. Although positively beneficial for sales statistics, continually replenished supplies mean that food stocks close to expiry are often overlooked by consumers (FAO 2011). This store's effort, however, resulted in a product display in store shelves that increased the amount of food wasted. Once the food is considered as waste, there are three different classification to manage organic waste: (a) stolen/distorted (products in good conditions but incomplete or altered due to thieves manipulation), (b) donation to those in need (products in good conditions but with aesthetically alterations in the package) and (c) food waste (rotten products, expiration date was passed, employees decided the landfilling of the food).

Two of them (stolen/distorted and donation) are regarded as second quality products as they are aimed to serve a different purpose in the food value chain. In this case, the destination of those second quality products is their distribution to multiple food charity organisations. In that sense, the Walmex store has been working on food donation especially with a national charity organisation that redistributes the food through their soup kitchens. Currently, they have been donating to this charity only packaged food with aesthetical alteration, placing aside vegetables, fruits, bakery and pastry. Since after the collecting time (11:00 am), food cannot be stored for the next day, it is then regarded as waste. This is mainly due to risks regarding food quality, cost of transport and preservation which also requires additional labour, storage, examining and monitoring efforts to redistribute the saved food.

However, the third category (food waste) has no declared purpose in the food value chain; in consequence it is directly landfilled without taking any benefit out of it. The 12 respondents mentioned that the food waste is managed by staff that follows technical criteria (texture, colour and odour) as well as subjective criteria. They discard products with no good appeal that are interpreted as low quality according to customers' demands. Once the organic products are taken away from the store shelves, a green bag is used to collect the food labelled as organic waste to be transported into a refrigerated room. This procedure takes place at least three times per day. After the collection phase of the food waste, its management passes to an external partner who is in charge of the daily disposal of the food waste in landfill sites.

According to the recorded information, two main groups of food waste at the organic end-of-life chain were identified: (1) fruits and vegetables and (2) bakery and pastry. Approximately 95% were fruits and vegetables mainly bananas, apples, watermelons, tomatoes and potatoes (see Fig. 9.1). On the other group, bakery and pastry were represented mainly by cakes, bread and crackers.

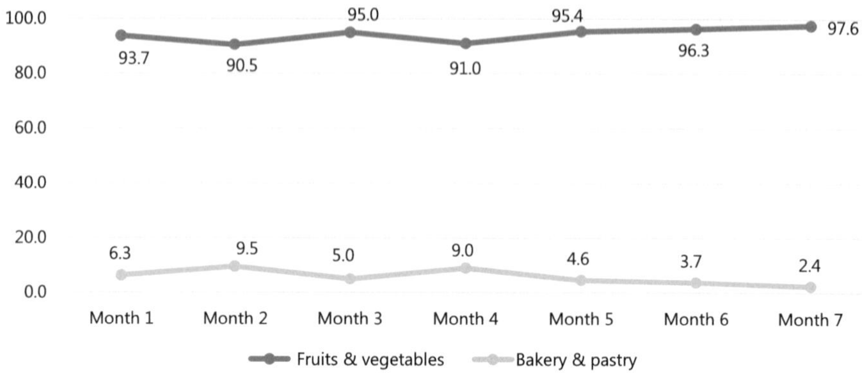

Fig. 9.1 Food waste landfilled per month (%)

As observed in Fig. 9.1, the daily amount of food discarded is not constant through the time. In fact, it can be stated that the gap between the lowest volumes landfilled and the highest volumes landfilled is significant. As indication, for fruits and vegetables in Month 2 (lowest volume of food waste landfilled) compared to Month 7 (highest volume of food waste landfilled), there is gap of more than 7%. The same occurred as to the group of bakery and pastry; the gap between the lowest volume landfilled (Month 7) and the highest volume landfilled (Month 2) was about 7%. There are some indications that this fluctuation might be caused by a poor supply chain management. However, the exact causes of having such fluctuation on the values were not mentioned during the interviews. Although, the proportion of food waste might not be significant during the retail operation; given the aim of this research, it is not only environmental and social wise to tackle this issue but also financially responsible to address the food waste problem as it was so far.

9.4.2 *ZWTL Strategy at Walmex's Store*

Prior to setting a ZWTL target, it is relevant to highlight all the possible scenarios that could reduce the food being landfilled. Also important to say is that these possible scenarios have to be firstly fully embedded into the environmental regulatory frameworks before any of them can be actually executed. In this case, a scenario was delivered prioritising food waste avoidance (reduce, feed people in need and feed livestock). The least desired option of the waste management approach corresponds to energy recovery, e.g. biodigester. This framework was completed according to the adaptation of the food waste hierarchy developed by Papargyropoulou et al. (2014). This provided a clear stance of the food conditions in the retail store, with the aim of creating a ZWTL landfill strategy under the food waste hierarchy principles. Therefore, we start with the most recommended one: reduction.

9.4.3 Reduction

At the retail store level, food waste can be attributed to two main factors: oversupply and reduction of shelf life. Those factors can be corrected by improving the supply chain management (Sciortino et al. 2016). At present some products in the store such as bananas, apples, watermelons, tomatoes, potatoes, cakes, bread and crackers have been identified as the products with less shelf life span. However, a data-driven supply network is required in the store to obtain more information about the shelf life of all entering products; this could ensure a better logistics approach to extend shelf life.

Consequently, the store's manager may then choose to sell food products based on their remaining shelf life. One major advantage of this strategy is that high-quality perishables can be sold to the customers at a premium cost by choosing the products with relatively long shelf lives (Jedermann et al. 2014). Upgrading these practices is important since employees' criteria to discard up food are ruled moderately by subjective standards as a decisive factor on whether to send the food or not to landfilling; by doing so, food stocks could be reduced as no high volumes of food are needed to be in store shelves. The 'first-expired, first-out' (FEFO) approach is one of the most applicable ways to deal with this issue. Although, FEFO is convenient for the store, it is important to mention that this approach is only possible by using robust inventories that enable accurate demand forecasting of those products (Hertog et al. 2014).

This approach is not only intended to treat with oversupply issues but also to an extent aimed to lengthen food shelf life. In the light of the benefits that this scenario might deliver, East (2011) proposed a system to predict storability performance testing FEFO compared to first-in, first-out (FIFO) and last-in, first-out inventory scheduling protocols. The author concluded that by using this approach, the improvement in storage capabilities is possible, so the reduction of fruit losses might rise up to 30% in comparison to other protocols. For more practical reasons, this research was focused in the postharvesting phase, especially at-harvest site. On the other hand, Sciortino et al. (2016) observed that 14% of food waste was reduced when decisions were made on a shelf-life-based analysis. This can be done by collecting accurate information about the daily flow of food, thus minimising food losses. Thereby, the volumes of food required to maintain the shelf full of high-quality food could be matched with costumers' demands and simultaneously cutting costs in storage, cooling and marketing (Sciortino et al. 2016).

9.4.4 Feed People in Need

At this point in time, there is no guarantee for the food stock to be completely sold out as the operations of forecasting and extending shelf life are also coupled with costumer's purchase behaviour which cannot be controlled. Hence, the 'fit-for-human-consumption' activity was enlisted as a 'second-best' solution in a context where there is still high-quality food in the retail store. Its purpose is to provide food to those unable to afford their own food. Thus, food banking emerged as non-governmental initiatives in the form of alliances saving food to distribute it to people in need (Schneider 2013). In conventional models of food banking, participants donate food that has little commercial value (overruns in production, excess supplies, past sell-by or stock that does not move on the shelves), but its nutritional properties remain for consumption purpose. At the retail level, operators can donate food that is too close to the expiring date losing its commercial value, or giving away unmarketable products such as bruised fruit and vegetables (Timmermans et al. 2014).

As a matter of fact, prior to this study, the store had previously defined its own target regarding food donation. The policy is set up to 30% of food losses and could be more as the managerial level considers it appropriate. Although, the target was set, an exact daily rate of food waste reduction after the previous phase (Reduction) is unknown. The Walmex's store must overcome potential barriers regarding logistics, information and legal concerns. Likewise, strengthening food banks and food bank networks throughout the country could be the sort of solution as a win-win situation that would benefit those in need without landfilling food. This situation implies less amount of food sent to donation but of a higher nutritional value due to the reduction-at-the-source strategy proposed for this study. So, it is worth to remark the relevance of a proper implementation of the food waste hierarchy used as a framework for this study, ever since if implemented properly, the amount of food meant to donation could be less but of at a higher nutritional value.

9.4.5 Feed Livestock

Another scenario of lesser value is to process food waste for animal feed. However, this is often not financially lucrative considering lower prices in that sector (MacArthur 2013). The livestock sector could use the retail flows of foods which cannot be redirected to human consumption either reduction through inventory management/extending shelf life span or food donations. The success of this scenario relies heavily on a number of practices, processes and policy parameters, such as safety, legislation on materials, operator rights and cost (Timmermans et al. 2014).

During the interview process, one person representing the livestock sector expressed his interest in the discarded food at the Walmex's store. According to this person, his sector may use for cattle-breeding up to 25% of bakery and pastry for the youngest group of animals. Then, this proportion could be raised up to 40% after 45 days of the animal lifetime. On the pig-breeding sector, after 25 days of the animal lifetime, their diet is a food mixture including among others bakery and pastry. After 45 days of the animal lifetime, fruits and vegetables conform part of their diet. Fruits and vegetables can be combined with bakery and pastry with portions between 35% and 15% of their diet. During the animal's adult life (e.g. pigs), the proportion of bakery and pastry can surpass 50% of the total amount of the food used to feed. The proportions on the diet were disclosed in order to indicate the livestock sector high demand of retail store's food losses which could close the loop of the ZWTL strategy.

As priory mentioned the livestock sector is highly interested in using the Walmex's food waste for cattle-breeding and pig-breeding. Indeed, the feeding method guarantees that fruits and vegetables are consumed immediately right after the food arrives to the farms. As for the food waste conformed by bakery and pastry, their intake was insured to a maximum of 3 days after their arrival. Even further in case the food might become not-fit-for-livestock consumption, the farms proved to have composting procedures for the no-longer-eatable food. Therefore, the farms themselves are not disposing of to landfill but rather closing the loop on food waste as well.

Setting quantifiable target for this scenario was not easy to define when comparing with the other two scenarios which are meant to close the loop on food waste. In any case, animal feeding remains as an option to achieve ZWTL through composting the remaining non-eatable portion. But due to the priority of reduction and donation streams, it seems irrelevant to fix a target for the feed livestock one. This latter depends on the quantity left after the two previous scenarios in the waste hierarchy. But in practise it is not a direct option for the remaining portion because this should comply with legislations associated to food safety and food quality standards at any time. Thus, not all the food is qualified to farm animals.

Even so, some numbers could be estimated aligning this strategy to the global target set by nations, regions and corporations on food waste avoidance. In that sense, the European Union and the United Nation target to decrease 50% of food waste (Lipinski et al. 2013). With regard to this scenario, feeding livestock as part of the food waste hierarchy framework might contribute up to 6% of the reduction once the two first scenarios have been exhausted (they account 44%) towards the ZWTL target (Fig. 9.2).

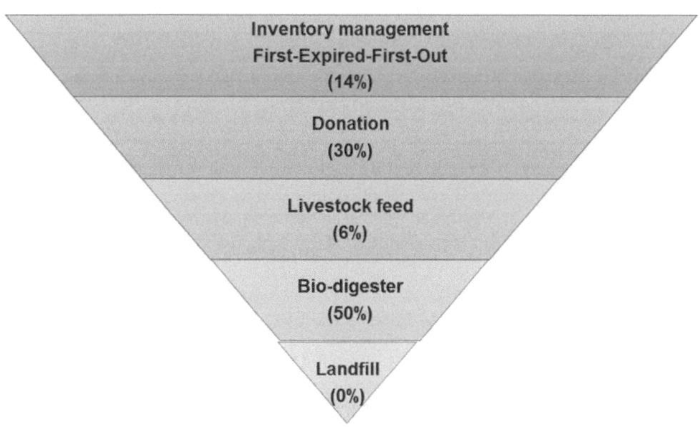

Fig. 9.2 ZWTL strategy. (Adapted from Papargyropoulou et al. (2014))

9.4.6 Biodigester and Energy Recovery

The final step to be undertaken, as it is illustrated in Fig. 9.2, is the recovering of energy by using a biogas system which appeared to be the suitable technology for this case study due to its advantages to treat food waste (MacArthur 2013). The nutritional richness of those materials is perfectly appropriate for any anaerobic digestion technologies to convert this influx into renewable energy and organic fertiliser (Leung and Wang 2016).

As explained here above, the three proposed scenarios must obey at least to a certain food safety standard for health reasons complying with the current Mexican legislation and ethical behaviour on social accountability. After all the three potential scenarios here studied, no additional option for reusing and recycling was feasible for some of the food becoming at that point waste, as no other option is feasible at present. Moreover, bearing in mind the inevitable biological decay of food quality, the organic waste must face the unavoidable act of being managed. This is because their natural decomposition creates odour, fast depletion on any physical surface, release of greenhouse gases and leachate percolation (Leung and Wang 2016). As observed in Fig. 9.2, 50% of the total food being disposed of by the retail store is available to be digested inasmuch as the former three scenarios have been already exhausted, and no other option is able to retain food quality and safety properties for either animal or human consumer any longer.

However, assuming that the remaining volume of food waste of the ZWTL scenario is still unknown, the biodigester system was designed by considering the complete daily amount of food wasted. From this assumption, the capabilities of the system were identified after some calculations. Furthermore, due to the existence of several Walmex stores in Mexico City, there is the possibility to bring their food waste to the biodigester system to reach the biodigester capacity.

The biodigester design can be daily fed by two groups of food waste (fruits and vegetable and bakery and pastry). The total substrate feeding the system was a water mix ratio of 1:3 which was calculated according to Arboleda and Salcedo (2009). Additionally, the volume substrate was estimated considering a density average of 1.3 kg/l at 35 Celsius degrees. After calculations were done, it was possible to estimate the methane production based on the theoretical value of 500 ml $CH4$/g VSS. The outcome of this theoretical digestion process yielded on average 11054.27 $CH4$ litres per day as observed in Table 9.1.

Theoretically, it is possible to say that all the amount of food wasted can be used to feed a biodigester to produce electricity. Yet care should be taken in the anaerobic digestion process as inhibition may occur when food waste is used as the main substrate. This inhibition is related to the disparity in the nutrient content of food waste. Therefore the biogas yield is affected by any variation in carbohydrate, protein and lipid content (Chiu and Lo 2016). Special attention is focused on the lipids and lignocellulosic concentration in food substrate. This could lead to process inhibition if food waste with high lipid content was digested (Zhang et al. 2014). In case of

Table 9.1 Biogas production from food waste food

Month	Volume substrate (l/day)	Kg VSS/day	CH_4 production (l/day)
Month 1	697.5	19.5	9730.1
Month 2	660.1	18.4	9208.8
Month 3	774.9	21.6	10810.3
Month 4	560.5	15.6	7819.4
Month 5	970.2	27.1	13534.8
Month 6	928.4	25.9	12951.4
Month 7	955.2	26.7	13325.0
Average			11054.27

excessive addition of lignocellulosic waste in the substrate mixture, this might result in poor biogas production (Chiu and Lo 2016).

According to the properties of the food wasted, there was no significant presence of lignocellulosic biomass (no rice proportion was characterised), and the lipid content coming from bakery and pastry represents only 5.8% of the total wasted (see Fig. 9.1). Hence, it is expected to have standard performance of the anaerobic digestion system since the ingoing food waste flow does not content representative number of components to inhibit the biogas production.

As priory outlined the conditions in anaerobic digestion made possible to use methane (CH_4) as a source to generate heat and electricity. From calculations it was possible to estimate the feasibility of using renewable energy to substitute (a) electricity currently supplied by the grid and (b) heat from natural gas consumption. As shown in Table 9.2, 394913.7 (KJ/day) is the total energy a biodigester was able to produce under similar conditions than this case study. This value is the result of the methane theoretically produced in anaerobic digestion (11.1 Nm^3/day) multiplied by the low heating value (LHW). This latter suggested by the low heating value (LHW), as suggested by Waldheim and Nilsson (2001). However, the energy requirement for pasteurisation was extracted from the total energy production since this energy requirement is a stage prior to the digestion process.

Once, the total energy available for electricity generation was estimated assuming a combined cycle gas turbine efficiency of 50% as recommended by Honorio (2003). This was done in order to determine the kilowatts per hour the system was capable to supply when substituting electricity provided by the grid. From those calculations, the electricity produced by anaerobic digestion can cover 0.8% of the total electricity consumed per month. This electricity share, from the biodigester, is neither technical nor economically feasible as it is not supplying significant electricity to the system, and therefore the payback time is foreseen as negative.

Furthermore, a second calculation was done taking into consideration the generation of biogas for heat and cooking utilities trying to find some economic value to the biodigester by-products. This resulted in 254.73 MJ/month of energy consumed for heat and cooking as observed in Table 9.2. This value was discounted of the total energy for electricity generation. This result considered an efficiency of heating of 80% (standard value) and an efficiency of combined cycle gas turbine of

Table 9.2 Energy recovery (power and heat) by biogas production

Power (electricity)	
CH_4 production (Nm^3/day)	11.1
Low heating value (MJ/ Nm^3 CH_4)	35.7
Total energy (KJ/day)	394913.7
Energy requirement for pasteurisation (KJ/day)	48512.2
Total energy for electricity generation (KJ/day)	346,401
Efficiency of combined cycle gas turbine (%)	50
Share in the total electricity consumption per month (%)	0.8
Heat (cooking)	
Higher heating value natural gas (MJ/Nm^3)	33.7
Energy consumed by heat and cooking (MJ/month)	254.7
Total energy for electricity generation (KJ/day)	346,401
Efficiency of heating (%)	80
Efficiency of combined cycle gas turbine (%)	50
Electricity produced by biogas after heat and cooking (KWh/month)	1399.1
Savings in energy bills (%)	18.6
Payback time (years)	2.4

50%. Thus, it resulted in 1399.1 KWh/month of electricity produced by biogas after heat and cooking.

The electricity produced by biogas after heat and cooking covers the energy requirements for those operations in the retail store, and there can still be an energy value left for electricity generation. However, it does not directly mean that this amount of energy would replace the electricity consumed from the grid, but rather it implies that biogas can substitute (commercial) natural gas for heat and cooking operations.

The biodigester technical feasibility was confirmed through cost-benefit analysis. According to this analysis, a commercial biogas plant (only bioreactor system for heat and cooking) at a cost of 66,000 USD with a cost operation per year of 3800 USD is stablished at site. Thus, the savings in energy (electricity and heat and cooking) bills per month could be lowered in 18.6% (Table 9.2). Therefore, the payback time for this type of solution will take 2.4 years.

9.5 Discussion

In this section it is described the combination between ZWTL strategy and the ReSOLVE framework as the selected approach to reach a CE business model.

9.5.1 The ReSOLVE Framework Towards a Circular Business Model

Approaches to close the loops on food waste at Walmex were here analysed by using the ZWTL strategy as framework. The ultimate target was indeed 0% of waste for disposal in landfills (Fig. 9.2). This served indeed as a showcase to identify those waste (materials) streams that could jointly lead to the 0% waste to landfill. But due to the multinational character of Walmex, this case has proved to feature some patterns that could be replicated beyond the Mexican context if those patters fit into international frameworks such as those published by the Ellen McArthur Foundation (2013). Hence, in this section, the technical proposal, so far here discussed, is in this section coupled to the ReSOLVE conceptual framework (MacArthur Foundation 2013). By integrating the ReSOLVE and ZWTL frameworks, technical, environmental and financial benefits can be analysed at once and serve as a tool to be explored in other Walmex's stores in Mexico or in similar food waste generation and composition conditions. The final purpose is to deliver a general model that can deliver a closer approach to circular business models by deploying closing loops of food materials.

9.5.2 Closed-Loop System

The business action in the ReSOLVE framework called Loop was chosen as the first scenario to be assessed towards a CE business model. This is due to the principle of the Loop concept and the CE itself meant to keep a closed loop and prioritise inner loops (MacArthur 2013). As outlined in Sect. 9.3, the current management in the Walmex's store disposed of a relevant amount of food waste in landfills. Therefore, by using this framework, a fully closed Loop in a ZWTL strategy is proposed (Fig. 9.3).

In this scenario, the biodigester was displayed as the main/unique treatment; thus, the Closed- Loop system is considered from a technical and economical perspective. Technical-wise, the system can cover all the heat and cooking demand for the store as calculations showed (Table 9.2), while it provides a source of renewable fuel (CH_4) to replace the fossil fuel. At the same time, this system is avoiding the emission of greenhouse gases in open landfills by combusting the methane in a biogas plant. However, this system is lacking fully environmental awareness as it is fostering the production on food waste to feed the system. Hence it contradicts, to some extent, the facts shown in Sect. 9.3, where the food still edible for human consumption was highly recommended to not be biodigested first. Furthermore, the financial benefits can be easily recognised as shown in Table 9.1; the savings in energy bills might be up to 18%.

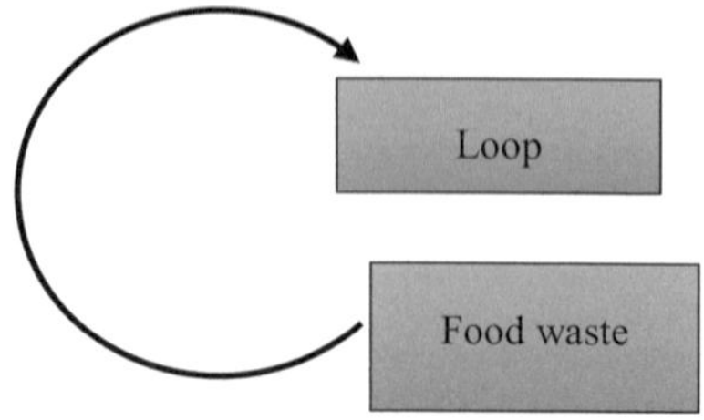

Fig. 9.3 Closed-loop system

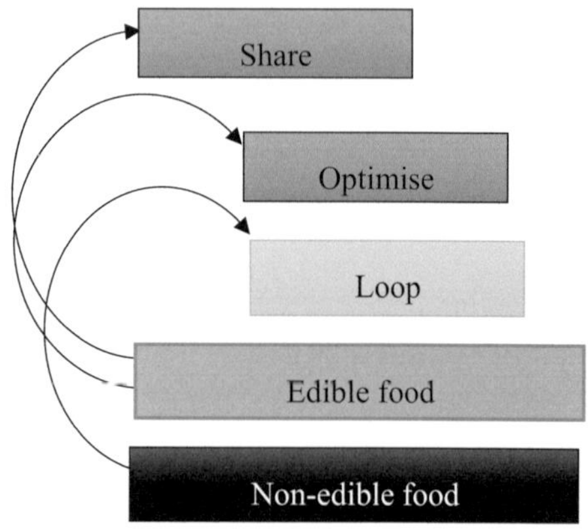

Fig. 9.4 Share, optimise and closed-loop system

9.5.3 Share, Optimise and Closed-Loop System

The Share, Optimise and Closed-Loop system does not follow the principles of CE of keeping a closed loop and prioritise inner loops as previously showed. Instead, it can be seen as an integrative approach by adapting the food waste hierarchy in the ReSOLVE framework (see Fig. 9.4). The share is meant for donations, but it can be referred to feeding livestock as well since both maximise product utilisation (MacArthur 2013). Optimise is related to the reduction phase as it removes waste by increasing efficiency (first expired, first out). And finally, the Loop referred to the anaerobic digestion process of recovering energy from the food waste stream.

From the technical standpoint, Share (donation and animal feeding) was accountable for approximately 36% of food waste reduction. Then, optimise could represent up to 14% food waste reduction. And lastly the Closed-Loop through the use of a biodigester represented 50% of the food waste stream (see Fig. 9.2). This technical system can also be regarded as a non-wasteful use of resources system, ever since it is focused in an intensive waste-reduction approach. The environmental externalities might be dramatically reduced as no food is wasted in landfills, the same as the Closed-Loop system. Albeit, the former system aimed to tackle food waste at its very primary stage, avoiding the necessity of generating food waste by cutting down the oversupplying of perishables at the Walmex's store.

On the financial aspect, the savings come from different sources. In Share, donation is seen as highly altruist, but it needs some compensation to be a suitable solution. Some fiscal incentives promote this initiative to compensate the economic losses in the business. Animal feeding although is not the ideal option for edible food; it also might recover some economic value while providing good-quality food to farm animals. In addition, increasing efficiency through Optimise (reduction) brings savings to the business.

When Share and Optimise are summed up, the savings reached 23% compared to the total cost of the food landfilled. The financial benefits were estimated taking into account the fiscal incentives for donating food, the cost of perishables for the livestock sector and the cost reduction by using first-expired, first-out approach. Nonetheless the savings can be higher by adding the energy bill savings in gas consumption through the use of the biodigester. Placing these elements together, the savings could be up to 41%. Yet, the biodigester system might need resources from other Walmex's stores, since in order to cover the heating and cooking demand, the system must work the closest to full capacity.

9.6 Conclusions and Recommendations

The main research question driving this research was: how the organic waste management in Walmex can be improved by combining a CE business model and a ZWTL strategy? To answer this question, the food waste hierarchy provided a suitable framework to target a ZWTL goal. It was noted that the current organic waste stream at the store has relevant opportunities to be optimised. The framework used showed that besides the clear oversupply issues diagnosed, most of the food can be recovered through different stages. Therefore, boosting managerial performance rather than implementing sophisticated technological developments on perishable foods seems feasible as this measure requires considerably involvement of a high number of staff members (e.g., operators should be skilled and knowledgeable) and low investment. Prior to implement any waste management action, the regulatory frameworks have to be considered to keep environmental compliance at all levels.

Businesses use different tools to assess their performance towards sustainability. Albeit, when it comes to specific sustainable issues, e.g. food waste, the literature is still scarce. Furthermore, a sustainable proposal must not only be effective but rather integrative due to its cross-cutting nature of sustainable practises. Therefore, for the purpose of this case study, several frameworks were integrated seeking for ZWTL strategies. In particular the ReSOLVE framework served to evaluate the CE principles which theoretically can be converted into six business actions. For this research three business actions were placed into consideration: Share, Optimise and (closed) Loop (SOL). Basing the analysis on two scenarios, the integrated system that appeared to be the most circular is the one using SOL. This system is clearly deep-rooted in the philosophy of the CE as it is handling the food waste by reducing the amount on nonedible organics. Thus, the SOL4FoodWaste framework is delivered as an attempt to collect applicable sustainable concepts and integrate them as one single framework that might potentially be implemented or resembled in similar retailer industries.

Although the key concepts and benefits of the ZWTL and CE principles have been available for many years, very often the technical-environment-financial relationship is misaligned that has resulted, so far, in not circular systems. The findings of the SOL4FoodWaste framework showed the benefits of managing the food waste through the three business actions, saving about 40% of the food waste management costs.

9.6.1 Recommendations for Further Research

Because in this study the main focus was the operations of food waste at one of the retailer stores of Walmex, the role of costumers on the ZWTL strategy was not analysed. Although, consumers' behaviour was not part of the study, scholars have agreed on consumers' power to influence product quality standards. There is a constant perception that consumers will not buy food which has the 'incorrect' dimension or poor appearance. At present there is weak existing factual information about consumer's preferences regarding the food appearance in connexion to quality and safety standards. Hence one of our recommendations for future research is to inquiry the consumers in order to explore the possibility to offer them a broader quality range of products in the retail stores not only based on good appeal. By influencing the costumers, a dramatic reduction of food losses can be foreseen. In addition to that, further research must be focused on verification procedures of the SOL4FoodWaste framework implementation. This type of integrative managerial approach can motivate the development of circular business models/actions for other type of discharged materials from the technical loops (MacArthur 2013), e.g. clothing, cleaning products and electronics, among others.

References

Adhikari BK, Barrington S, Martinez J (2006) Predicted growth of world urban food waste and methane production. Waste Manag Res 24(5):421–433

Adhikari, B. K., Barrington, S. F., & Martinez, J. (2009). Urban food waste generation: challenges and opportunities. Int J Environ Waste Manag, 3(1–2), 4–21.

Arboleda Y, Salcedo L (2009) Fundamentos para el diseño de Biodigestores. Universidad Nacional de Colombia, Palmira

Blackburn J, Scudder G (2009) Supply chain strategies for perishable products: the case of fresh produce. Prod Oper Manag 18(2):129–137

Bond, M., Meacham, T., Bhunnoo, R., & Benton, T. (2013). Food waste within global food systems. Global Food Security Programme. A global food security report. Swindon

Calabro PS (2009) Greenhouse gases emission from municipal waste management: the role of separate collection. Waste Manag 29(7):2178–2187

Chiu SL, Lo IM (2016) Reviewing the anaerobic digestion and co-digestion process of food waste from the perspectives on biogas production performance and environmental impacts. Environ Sci Pollut Res:1–16

Chiu SL, Lo IM, Woon KS, Yan DY (2016) Life cycle assessment of waste treatment strategy for sewage sludge and food waste in Macau: perspectives on environmental and energy production performance. Int J Life Cycle Assess 21(2):176–189
Colon M, Fawcett B (2006) Community-based household waste management: Lessons learnt from EXNORA's 'zero waste management' scheme in two South Indian cities. Habitat Int 30(4):916–931
East A (2011) Accelerated libraries to inform batch sale scheduling and reduce postharvest losses of seasonal fresh produce. Biosyst Eng 109(1):1–9
FAO (2011) Global food losses and food waste—extent, causes and prevention. UN Food and Agriculture Organization, Rome
Forster P, Ramaswamy V, Artaxo P, Berntsen T, Betts R, Fahey DW Myhre G (2007) Chapter 2: Changes in atmospheric constituents and in radiative forcing. In: Climate change 2007. The physical science basis
Ghisellini P, Cialani C, Ulgiati S (2016) A review on circular economy: The expected transition to a balanced interplay of environmental and economic systems. J Clean Prod 114:11–32
Hertog ML, Uysal I, McCarthy U, Verlinden BM, Nicolaï BM (2014) Shelf life modelling for first-expired-first-out warehouse management. Philos Trans A Math Phys Eng Sci 372(2017):20130306
Honorio L (2003). Efficiency in electricity generation. Eurelectric: Union of the Electric Industry. VGB Powertech, Brussels
Jamasb T, Nepal R (2010) Issues and options in waste management: A social cost–benefit analysis of waste-to-energy in the UK. Resour Conserv Recycl 54(12):1341–1352
Jedermann R, Nicometo M, Uysal I, Lang W (2014) Reducing food losses by intelligent food logistics. Philoso Trans A Math Phys Eng Sci 372(2017):20130302
Jurgilevich A, Birge T, Kentala-Lehtonen J, Korhonen-Kurki K, Pietikäinen J, Saikku L, Schösler H (2016) Transition towards circular economy in the food system. Sustainability 8(1):69
Kim M-H, Song Y-E, Song H-B, Kim J-W, Hwang S-J (2011) Evaluation of food waste disposal options by LCC analysis from the perspective of global warming: Jungnang case, South Korea. Waste Manag 31(9):2112–2120
Koester U (2015) Reduction of food loss and waste: an exaggerated agitation. EuroChoices 14(3):34–38
Kothari R, Kumar V, Panwar N, Tyagi V (2014) Municipal solid-waste management strategies for renewable energy options. In: Sustainable bioenergy production. CRC Press, Boca Raton, pp 263–282
Leung DY, Wang J (2016) An overview on biogas generation from anaerobic digestion of food waste. Int J Green Energy 13(2):119–131
Lewandowski M (2016) Designing the business models for circular economy—towards the conceptual framework. Sustainability 8(1):43
Lin CSK, Pfaltzgraff LA, Herrero-Davila L, Mubofu EB, Abderrahim S, Clark JH, Dickson F (2013) Food waste as a valuable resource for the production of chemicals, materials and fuels. Current situation and global perspective. Energy Environ Sci 6(2):426–464
Lipinski B, Hanson C, Lomax J, Kitinoja L, Waite R, Searchinger T (2013) Reducing food loss and waste. World Resources Institute Working Paper, June
MacArthur E (2013) Towards the circular economy. J Ind Ecol
Mason I, Brooking A, Oberender A, Harford J, Horsley P (2003) Implementation of a zero waste program at a university campus. Resour Conserv Recycl 38(4):257–269
Mohee R, Mauthoor S, Bundhoo ZM, Somaroo G, Soobhany N, Gunasee S (2015) Current status of solid waste management in small island developing states: a review. Waste Manag 43:539–549
Møller J, Boldrin A, Christensen TH (2009) Anaerobic digestion and digestate use: accounting of greenhouse gases and global warming contribution. Waste Manage Res 27(8):813–824
Moumen A, Azizi G, Chekroun KB, Baghour M (2016) The effects of livestock methane emission on the global warming: a review. Int J Glob Warm 9(2):229–253

Müller C (2007) Anaerobic digestion of biodegradable solid waste in low-and middle-income countries. Sandec report
Murphy S, Pincetl S (2013) Zero waste in Los Angeles: Is the emperor wearing any clothes? Resour Conserv Recycl 81:40–51
Murray A, Skene K, Haynes K (2015) The circular economy: an interdisciplinary exploration of the concept and application in a global context. J Bus Ethics:1–12
Papargyropoulou E, Lozano R, Steinberger JK, Wright N, Bin Ujang Z (2014) The food waste hierarchy as a framework for the management of food surplus and food waste. J Clean Prod 76:106–115
Parfitt J, Barthel M, Macnaughton S (2010) Food waste within food supply chains: quantification and potential for change to 2050. Philoso Trans R Soc Lond B 365(1554):3065–3081
Parliament E (2008) Directive 2008/98/EC of the European Parliament and of the Council on waste and repealing certain Directives. Off J Eur Union L 312:3–30
Pitt J, Heinemeyer C (2015) Introducing ideas of a circular economy. In: Environment, ethics and cultures. Springer, Berlin, pp 245–260
Planing P (2015) Business model innovation in a circular economy reasons for non-acceptance of circular business models. Open J Bus Model Innov
Ray N, Mohanty M, Mohanty R (2013) Anaerobic digestion of kitchen wastes: "biogas production and pretreatment of wastes, a review". Int J Sci Res Pub:286
Schneider F (2013) Review of food waste prevention on an international level. Paper presented at the Proceedings of the Institution of Civil Engineers-Waste and Resource Management
Sciortino R, Micale R, Enea M, La Scalia G (2016) A webGIS-based system for real time shelf life prediction. Comput Electron Agric 127:451–459
SEMARNAT (2015, May 12) Secretaría de medio ambiente y recusos naturales. Retrieved from Secretaría de medio ambiente y recusos naturales: http://www.semarnat.gob.mx/temas/residuos-solidos-urbanos. Accessed 21 Aug 2016
Song Q, Li J, Zeng X (2015) Minimizing the increasing solid waste through zero waste strategy. J Clean Prod 104:199–210
Thi NBD, Kumar G, Lin C-Y (2015) An overview of food waste management in developing countries: current status and future perspective. J Environ Manag 157:220–229
Timmermans A, Ambuko J, Belik W, Huang J (2014) Food losses and waste in the context of sustainable food systems. HLPE, Rome
Villanueva A, Wenzel H (2007) Paper waste–recycling, incineration or landfilling? A review of existing life cycle assessments. Waste Manag 27(8):S29–S46
Waldheim L, Nilsson T (2001) Heating value of gases from biomass gasification. Report prepared for: IEA bioenergy agreement, Task, 20
Walmart (2015, Diciembre 31) Walmex. http://walmart16.ciro17mx.net/pdf/esp/walmart_2015_informe_anual.pdf. Accessed 25 Aug 2016
Wang L-H, Wang Q, Cai W, Sun X (2012) Influence of mixing proportion on the solid-state anaerobic co-digestion of distiller's grains and food waste. Biosyst Eng 112(2):130–137
Zaman AU, Lehmann S (2013) The zero waste index: a performance measurement tool for waste management systems in a 'zero waste city'. J Clean Prod 50:123–132
Zhang C, Su H, Baeyens J, Tan T (2014) Reviewing the anaerobic digestion of food waste for biogas production. Renew Sust Energ Rev 38:383–392

Chapter 10
PetStar PET Bottle-to-Bottle Recycling System, a Zero-Waste Circular Economy Business Model

Jaime Cámara-Creixell and Carlos Scheel-Mayenberger

Abstract The lack of environmental awareness in society, especially in developing countries, combined with inefficient waste handling systems, has caused millions of PET bottles to end up in landfills, losing their original value. In the worst cases, the bottles mishandled by consumers enter natural systems generating significant negative externalities such as the pollution of soil and water, with the possibility of reaching the oceans. In general, the plastic recycling industry in undeveloped economies is highly dependent on the participation of a broad social group known as scavengers, whose role is very valuable for the supply chain, although it is performed under very difficult conditions and usually operates as part of the informal economy. Maintaining the sustainability of the three actors involved in the industry—the environment, society and business—requires a different business model in which all actors must participate and produce a more inclusive added value. PetStar is a company that has designed and implemented a circular economy business model for PET bottles that is trying to achieve a dream for the recycling industry: to disengage the recycled bottle from virgin resources, avoiding the conversion of the packaging to waste, and operate a perennial cycle in the use of the packaging. The PetStar sustainable business model is presented, explaining how it works and how it creates a sustainable cycle that is economically feasible and competitive, environmentally resilient and socially shared among one of the poorest and most informal sectors, the scavenger and collector communities. This paper discusses a model of sustainable recycling that can be replicable in other developing countries where the same problem emerges, as well as how to generate a zero-waste circular value system that can be inclusive, clean, viable and capable of generating shared wealth for the community.

J. Cámara-Creixell (✉)
PetStar, Toluca de Lerdo, Mexico
e-mail: jcamara@petstar.mx

C. Scheel-Mayenberger
Instituto Tecnológico de Estudios Superiores de Monterrey, EGADE Business School, Monterrey, NL, Mexico
e-mail: cscheel@itesm.mx

M.-L. Franco-García et al. (eds.), *Towards Zero Waste*, Greening of Industry Networks Studies 6, https://doi.org/10.1007/978-3-319-92931-6_10

Keywords Zero-waste to landfill · Circular economy · Zero-plastic leakage · PET bottles · Plastic industry · Developing countries · Sustainable business model · Bottle-to-bottle recycling · Scavengers · Carbon footprint · Water footprint · Renewable energy

10.1 Introduction

Due to its functionality, resistance, safety and lightweight, the use of PET (polyethylene terephthalate) as packaging (bottles) in the beverage industry boomed in the 1990s, creating an environmental challenge because those bottles need to be responsibly disposed of by the consumer for proper collection and recycling.

Unfortunately, society's lack of environmental awareness, especially in developing countries, combined with inefficient waste handling systems, has caused millions of bottles to end up in landfills and lose their original value. In the worst cases, the bottles mishandled by consumers enter natural systems generating significant negative externalities, such as the pollution of soil and water, with the possibility of entering the oceans.

Based on these assumptions—is it possible to develop a way to maintain a PET bottle inserted into a permanent circular system, achieving zero waste and an economic viable business model, that creates social and environmental values?

To address this challenge in Mexico, PetStar started in 1995 as a collector of post-consumer PET bottles, installing a broad nationwide collection infrastructure, and has been buying the PET bottles in bulk format directly from the scavengers. At that time, most of the PET bottles were recycled and transformed into polyester fibres. The material collected was transported to PetStar's collection plants where it was hand-sorted in conveyor belts and then baled to be shipped to customers in the United States and China. Back then, there was no recycling infrastructure available domestically.

It was not until 2000 that PetStar was exposed to new technologies, which were mainly developed in Europe. Those technologies allowed recycled PET resin to be returned to the bottle production factories to be incorporated in new food-grade bottles within a closed continuum cycle, as recycled content.

With the available technologies approved by the FDA (Food and Drug Administration), PetStar tested balled bottles collected in Mexico that were dirtier than those collected in Europe and came up with the most suitable technologies for the project: (a) an Italian one for grinding and washing and (b) a Swiss one for extrusion and solid stating.

Once the collection infrastructure was developed and the technologies were chosen, the subsequent step was to identify the investors. In 2006 investors were found and the bottle-to-bottle project started. After securing customers and developing the engineering and construction, in 2009 the production of Phase 1 started, with a capacity to produce 20,000 ton/year of food-grade recycled PET resin.

To comply with its commitment with the environment by integrating recycled content into their new PET bottles, in 2011 Arca Continental (49.9% of share), the second largest Coca-Cola bottler in Latin America, bought PetStar. The group of shareholders grew when other Coca-Cola Mexican bottling companies joined Arca Continental, including *Coca-Cola de Mexico* (30%), *Bepensa Bebidas* (10%), *Corporacion del Fuerte* (5%), *Corporación Rica* (2.2%), *Grupo Embotellador Nayar* (1.9%) and *Embotelladora de Colima* (1%). Their shareholding structure is proportional to their market share, so by contractually committing to buy their proportion of food-grade recycled PET resin produced by PetStar, all end up having the same PET recycled content in their new bottles.

The new owners, all part of the Mexican Coca-Cola Industry (MCCI), invested in Phase 2 to expand the capacity to recycle 3.1 billion bottles per year into 50,000 tons of high-quality PET food-grade recycled resin. Production started late in 2013, and PetStar was nominated as "The World's Largest Food Grade PET Recycling Plant" by "PCI PET Packaging Resin and Recycling, Ltd." (McGeough 2013).

In the World Economic Forum 2016 (WEF-2016) in Davos, Switzerland, the Ellen MacArthur Foundation (EMF) launched its study "The New Plastics Economy– Rethinking the Future of Plastics"(MacArthur et al. 2016). This was announced with the intention of offering a new vision aligned with the circular economy concept (Webster 2015; EMF 2013) and capturing the current opportunities, as only 14% of the global plastic packaging produced was collected for recycling, from which just 2% was recycled into the same or similar quality application. Ninety-five per cent of the material value ($80–120 billion/year) of plastic packaging is lost after a short first usage. This situation, in addition to the leakage effects of non-recycled materials, has been causing degradation of the environmental system. Without significant action, by 2050 the ocean may contain more plastic than fish (by weight).

In response to this important challenge, in 2016 PetStar launched its PetStar sustainable business model (PSBM), reflecting its efforts to become a front-runner and innovative engine in the plastic recycling sector. The PSBM is fully integrated into the value chain of PET bottles, from the direct collection of post-consumer bottles (with scavengers and collectors) all the way to the production of new PET bottles combining virgin raw materials with recycled content. Simultaneously, the PSBM considers circularity in its core business because it promotes local job creation, technological innovation and new products that respond to local markets and engages community members in its operation.

The WEF-2017 was the platform for the EMF to launch a second study: "The New Plastics Economy-Catalysing Action" (MacArthur et al. 2017). In this version, it was mentioned that a global momentum for a fundamental plastic rethink is greater than ever. This report is the first to provide a concrete set of actions to drive the transition from the current 14% of plastic packaging collected globally for recycling to 70%. Such a transition was based on three strategies differentiated per market segment: (1) without fundamental redesign and innovation, about 30% of plastic packaging will never be recycled, (2) for at least 20% of plastic packaging reuse provides an economically attractive opportunity[1], and (3) with concerted efforts in

[1] In 2016, the MCCI sold 22.5% of its volume in returnable (reusable) PET bottles.

design and after-use processes, recycling would be economically attractive for the remaining 50% of plastic packaging.

The PSBM implemented during 2016 fully matched with six out of a total of seven premises of the catalysing action report (MacArthur et al. 2017). The premise not addressed by the PSBM corresponds to plastic films that do not apply to PetStar's profile.

The PSBM focuses mostly on the strategy (3) as the best approach because it covers both economic attractiveness and an efficient recycling system for PET bottles in closed-circled loops.[2]

This chapter is organised as follows: The first two sections focus on framing the situation of the PET industry in the world and in Mexico. The following section presents a detailed description of the PetStar sustainable business model and the processes the company has implemented to attain international standards and certifications, as well as a high degree of viability on the circularity of its operations. The last section deals with the benefits that the company has been produced to the society, to the environment, as well as to their shareholders. This chapter concludes with some insights about the importance to implement an optimum closed circular bottle-to-bottle cycle, maximising the environmental value of the material.

10.2 The Circular Economy Model of the PET Industry

The rethinking of the plastic industry requires a new approach to capture new opportunities based on circularity, new technologies and business models, especially for the packaging subsector (The EMF 2013, 2017).

As mentioned before, a great opportunity lies in how to recover, transform and reuse the large volumes of PET material that today are lost after a short first use and how to make this a profitable and sustainable industry.

Globally the PET used in beverage bottles has a higher recycling rate than any other type of plastic, "but unexpectedly, almost 50% of these containers are not collected for recycling and only 7% is recycled in a closed loop of bottle-to-bottle"; this represents a great opportunity.

What's more, the same study shows that "72% of plastic packaging is not recovered at all". Forty per cent goes to landfill, most often in deplorable conditions of pollution, health and environmental care; and 32% leaks out of the collection system and mostly goes to illegal dumping and/or mismanaged actions.

Finally, 14% of the plastic is sent to an incineration or energy recovering process. Although this is better than dumping, "the process loses the embedded effort, labour and costs needed to produce the finished product, and an over-deployment of this process may create a lock-in effect that may push higher value mechanisms (such as recycling) out of the packaging market".

---

[2] In 2016, the PetStar shareholders incorporated an average of 32% of PET recycled content in their bottles, with most of the green bottles at 100%.

All four of these situations are a great and irreversible threat for the planet and need to be converted into opportunities, mainly for developing countries, where these figures are rather dramatic.

In summary, some of the actions that must be taken include:

(a) How to increment the 14% of the plastic that is collected for recycling
(b) How to increment the 7% of the PET recycled and insert it into a bottle-to-bottle process
(c) How to take advantage of the 72% that is not recovered and is dumped into polluted landfills and where the collection system can be improved substantially
(d) How to reduce the 14% of the plastic sent to incineration and transform it into more valuable resources
(e) How to redesign the packaging so that it complies with the guidelines for plastics recyclability and does not contain components that are expensive to separate or render the container nonrecyclable
(f) How to reduce the negative externality of the low costs of oil and, at the same time, decouple plastics from fossil feedstock

Most of the above actions can be solved if a circular economy approach can be applied to convert negative practices into valuable opportunities for firms belonging to the important subsector of packaging in the plastic industry.

10.3 The Post-consumer Management of the PET Bottles in Mexico

Historically there was a void in the legal provisions in Mexico for the management of solid waste, its antecedent existing in the "Ley General del Equilibrio Ecológico y Protección al Ambiente" (General Law of Ecological Equilibrium and Protection to the Environment) as recently as 1988, and it was not until 2003 that the "Ley General para la Prevención y Gestión Integral de los Residuos" (General Law for the Prevention and Integral Management of Residues) was published; however, it considered solid waste from a perspective that led only to the disposal of waste generated by people or businesses and assigned local governments the responsibility of offering clean-up services or the option to award concessions for it. This scenario resulted in the appearance of multiple dumps (many uncontrolled), as well as landfills, where the logic has been to bury all types of waste, both recoverable and nonrecoverable.

Without an adequate legal framework, for decades the integrated attention to the waste and the recycling culture ceased to be promoted, a situation that today persists in most of the population and institutions in Mexico. The separation of recoverable waste from the source was not promoted nor was the shared responsibility addressed until 2006 when changes were introduced to the law in this matter to emphasise shared responsibility and the possibility of valuing the waste to take advantage of it through recycling (SEMARNAT 2006). Nevertheless, in Mexico it is possible to find valuable efforts to change this reality, as in the case of PET plastic packaging.

In 2002, when the PET collection rate and the recycling infrastructure in Mexico were almost non-existent, the Mexican bottling industry, seeking to increase collection rates to change the path of the chicken and the egg, created and funded "Ecología y Compromiso Empresarial" (Ecology and Industry Commitment) ECOCE (2017) as a nonprofit organisation to: (1) promote and fund the collection of post-consumer PET bottles to create certainty and develop the domestic recycling infrastructure; (2) develop media campaigns and environmental educational programmes to create awareness in consumers and promote better habits for the disposal of residue in society; (3) encourage synergies with different levels of the government to attend specific critical areas of mishandled PET residues, seeking solutions under the criteria of integral handling of solid residues by attacking the root cause; (4) participate and promote legislation and normativity related to solid residues in the waste stream to create legal certainty and viability for domestic collection and recycling; and (5) promote the development of a formal strong recycling industry in Mexico to encourage a domestic market, create formal jobs and establish certainty in the consumption of the material.

In 15 years of existence, ECOCE has been a key player in the development of formal PET recycling in Mexico. In 2016, 425,000 tons of PET bottles were collected representing a 57% collection rate of the 745,000 tons of bottles produced. With a total investment of 339 million dollars, the domestic recycling industry processed 240,000 tons that represent 56.3% of the collected bottles, while the remaining 43.7% was exported mainly to the United States and China.

10.4 The PetStar Sustainable Business Model

PetStar's operation generates a triple positive impact on the social, environmental and economic subsystems of the planet's natural ecosystem, which is 100% internalised in Mexico and embedded in the PetStar sustainable business model (PSBM).

In the environmental subsystem, in addition to removing the bottles from the surroundings, it contributes to a cleaner environment because, in the process of converting them into PET food-grade recycled resin, it reduces by 78% the greenhouse gas emissions against virgin resin. Since 2016, with a cogeneration project and the use of wind power, it was increased to 87% reduction (PetStar 2016), which is equivalent to removing 3.7 million cars from Mexico City for about 2 days, as shown in Table 10.1. This table presents the impact PetStar has delivered on the mitigation of recycled bottles versus bottles from virgin resources on tons of CO_2, as well as the percentage the company has been able to recycle and transform.

The positive social impacts are considered in terms of the provision of fair and stable income rates that dignify the conditions of the scavengers who indirectly

Table 10.1 Amount of CO_2 emission mitigation of Mexico City

Resin	**Emissions (tCO_2e/ton)**	**Production (ton/year)**	**Emissions (tCO_2e)**	**Mitigation vs Virgin (tCO_2e)**	**Mitigation vs Virgin (%)**
Virgin	2.330	50,000	116,500		
PetStar 2015	0.661	50,000	33,050	−83,450	72%
PetStar 2016	0.303	50,000	15,150	−101,350	87%
Cars in Mexico City (for PetStar Resin 2016)					
Mitigation vs virgin	101,350	tCO_2e			
Emissions per car in a year	4.80	tCO_2e			
Avoided cars per year	21,115	Cars			
Avoided cars per day	7,706,823	Cars			
Cars circulating per day	3,693,351	Cars			
Mitigated days	2.1	Days			
Sources					
Emissions (tCO2e/ton)	Assessment of PetStar's Sustainability Performance in 2016				
	PCI Wood Mackenzie by Helen MacGeough and Pieterjan Van Uytvank				
Emissions per car in a year	"Programa para mejorar la calidad del aire de la Zona Metropolitana de la Ciudad de Mexico 2011–2020" (2011–2020 programme to improve the air quality on the metropolitan area of Mexico City of Mexico City)				
	SEMARNAT y Secretaria de Salud (CDMX y Estado de Mexico)				
Cars circulating per day	"Programa para mejorar la calidad del aire de la Zona Metropolitana de la Ciudad de Mexico 2011-2020" (2011–2020 programme to improve the air quality on the metropolitan area of Mexico City)				
	SEMARNAT y Secretaria de Salud (CDMX y Estado de Mexico)				

work with PetStar through its multiple collection partners. As part of the social commitment towards this community, the company provides training and inclusion initiatives that result in certainty for a highly vulnerable population (Medina and Smith 2013).

One of those initiatives is the project developed in alliance with other civil society organisations (CSO) in the municipality of Chimalhuacan, in the state of Mexico. The project participants created the Integral Development Community Center (IDCC) that enhances opportunities with access to education, food and medical services for 250 children of the scavengers who live in the Tlatel Xochitenco community. The IDCC's purpose is to provide integral attention in the three fields mentioned above to build and develop life skills into the framework of childhood rights.

For educational purposes and as part of its social responsibility commitment, PetStar installed a Museum Auditorium whose mission is to promote environmental awareness, a recycling culture and the spread of the concept of shared responsibility among society, authorities and the private sector (Porter and Kramer 2011). Every year it receives, free of charge, around 13,000 visitors.

This space is environment friendly, with a system for collecting and treating rainwater, solar panels and a green roof, among other practices that help the environment, and is LEED Platinum certified[3] as a green building, making it the first museum in Latin America to achieve this level of certification.

At the economic subsystem, PetStar is part of the MCCI supply chain. The bottle recycling operation adds value and contributes to the sustainability of the packaging sector. With eight collection plants, one by-product valorisation plant, one recycling plant and multiple collection partners nationwide, PetStar directly employs almost 1000 people and indirectly economic benefits over 24,000 scavengers and collectors.

In congruence with its philosophy of excellence and in an effort to help Mexico comply with the Sustainable Development Goals (SDG) (United Nations 2015), PetStar joined the United Nations Global Compact and the Earth Charter and has the following certifications: ISO 9001, ISO 14001, ISO 22000, ISO 50001, OHSAS 18001, Clean Industry, Socially Responsible Company, Top Companies, Operation Clean Sweep, Clean Transport and Occupational Health and Safety Self-Management Programme.

10.4.1 The Circularity of the PSBM

The PetStar sustainable business model (PSBM) shown in Fig. 10.1 describes how the structure of PetStar is engaged in the sustainable PET packaging industry. The sustainability principle[4] includes most of the concepts that are embedded by the circular economy mechanisms of the ReSOLVE[5] framework.

From the above PSBM, the company manages four core processes: (a) an inclusive collection system, (b) the collection process, (c) the recycling process and (d) the marketing process. These processes are articulated through a circular economy framework that links all stakeholders with the main objective to attain a zero-plastic leakage and a sustainable region. A detailed description of each process follows.

[3] Leadership in Energy and Environmental Design.

[4] PetStar's Sustainability Report (2016).

[5] The ReSOLVE framework (regenerate, share, optimise, loop, virtualise, exchange) offers organisations a tool for creating value with circular approaches. Ellen MacArthur Foundation for Circular Economy, UK, 2013.

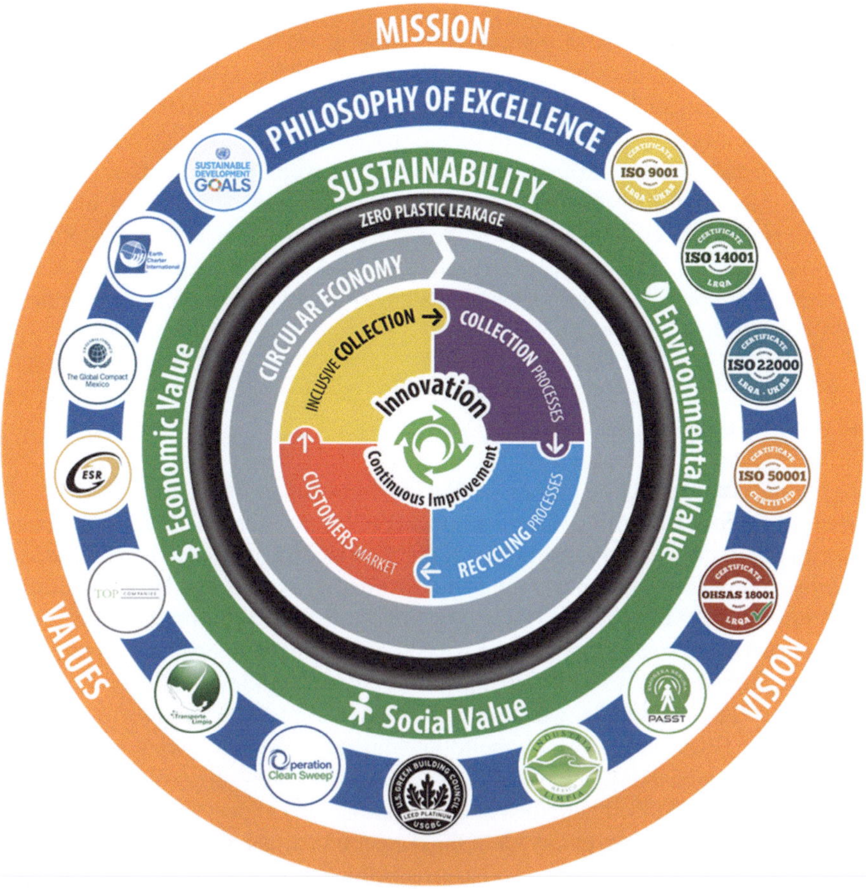

Fig. 10.1 PetStar sustainable business model (PSBM)

10.4.1.1 Inclusive Collection

PetStar promotes the concept of inclusive collection by participating in the Regional Initiative for Inclusive Recycling (RIIR), which is the main regional platform for the articulation of actions, investment and knowledge with the objective of improving scavengers' access to the formal recycling market in Latin America and the Caribbean by designing and implementing activities which allow (1) *improving* the socioeconomic situation of scavengers, (2) *facilitating* their access to the formal recycling market and (3) *promoting* the creation of public policies for integrated waste management which include scavengers (IRR 2011).

Through these actions, the RIIR would like scavengers' work to be recognised and valued by building a platform for multi-sector strategic alliances. These alliances will work as a space for dialog and action, thus increasing coordination capacity at the natural and regional level between the government, the business sector and the recycling organisations.

When the bottles are discarded by consumers, they consequently turn into residues. Those residues start a journey within the diverse solid waste management system locally available. Eventually, in some part of the system, they might be recovered by someone (mainly scavengers or collectors), either in their course (e.g. in a garbage truck, in the streets, in public spaces or at home) or at their final disposal site/landfill. Once the recyclable residue is recovered, the scavenger or collector has the option to sell it to a basic consolidator who, as its name indicates, consolidates the recyclables and commercialises them in the formal or informal recycling markets. This process takes place in many areas and at different scales, depending on the consolidator's location along the network. In Mexico, the rate of PET collection is 57% (ECOCE 2017).

10.4.1.2 Collection Processes

To optimise the collection processes, PetStar's collection division recently developed the PetStar inclusive collection model (PICM) that consists of nine processes that reinforce the robust infrastructure of eight collection plants and one by-product valorisation plant that are geographically distributed at a national level. This infrastructure includes the company's own specialised fleet of 120 trucks to offer a collection service of excellence to almost 1200 basic consolidators[6] called collection partners, offering certainty to around 24,000 scavengers and collectors who are recovering the bottles daily, which translates into a high social value.

The nine PICM processes are the following: (1) collection market intelligence, (2) prospection of the basic consolidator, (3) registration of the collection partner, (4) impeccable recollection to the collection partner, (5) impeccable payment to the collection partner, (6) development of the collection partner, (7) public policies, (8) social entrepreneurship and (9) service satisfaction to the collection partner.

All of the material recycled by PetStar is collected in bulk (loose bottles) format by its collection division, transported with its own bulk trucks to any of the eight collection plants where it is manually sorted by colour, removing some contaminants, and then baled. Eighty per cent of the total volume corresponds to the natural PET colour that is transported to the PetStar recycling plant in Toluca. The remaining 20% of other colours and some other types of plastics are classified in five different streams and transported to the by-product valorisation plant. The latter is also located in Toluca, where material is processed for other applications.

With the intention of demonstrating that recycling can be done in a responsible way (e.g. preventing ocean contamination), the leakage of loose bottles is avoided, from their transportation on the PetStar trucks all the way to the incorporation of the recycled resin in new bottles. Moreover, plastic leakage of bottles, caps, labels, flakes, dust and pellets is prevented.

[6] Information included in PetStar Internal Report (2016).

10.4.1.3 Recycling Processes

All the natural colour PET bales produced by the collection division are delivered to the recycling division that consists of one recycling plant located in the city of Toluca, state of Mexico. The recycling plant has the capacity to convert 65,000 tons per year of bales into 50,000 tons of food-grade recycled PET resin to produce new bottles with recycled content for the MICC. The amount indicated is equivalent to 3.1 billion bottles that could fill 2.4 times the Azteca Soccer Stadium[7] in Mexico.

The recycling processes in the recycling division are very sophisticated because they use state-of-the-art technology. The recycling division consists of three subprocesses: (1) grinding and washing, where the bales are broken and the loose bottles are fed into a very big bin with a conveyor belt that feeds a whole bottle prewash system. Afterwards the bottles are electronically sorted to remove colours and other polymers. When the material arrives at the grinding section, the bottles are cut into small pieces or flakes and go into the float sink tank, where the caps and labels will float while the PET material sinks. As the last operation in this process, they continue into friction washing, where glue and dirt are removed and, finally, the flakes are rinsed and dried to be stored in silos.

The second process corresponds to (2) *extrusion*: the washed flake is fed from the storage silos into the ring extruder where the material is melted, filtered and vacuumed. This latter phase has the purpose of removing humidity and volatiles (purification process) to finally produce amorphous pellets which are translucent, as the molecules are not aligned or disordered.

The third process is called (3) *solid stating*. In this process, the amorphous pellets are conveyed to the top of the 50-meter-high tower where the post-condensation or solid-stating process will take place. In this stage, temperature, pressure and residence time are controlled to provoke the re-polymerisation process. The latter will lengthen and align the molecules, building physical and thermal structure back into the resin, which will block the light, making it opaque. Additionally, the material will finalise its purification to become food-grade recycled PET resin.[8] The final product is stored in silos to await delivery to PetStar's customers.

As part of the recycling process, a series of by-products are generated, like the washed polyolefin flake of caps and labels, and different sorts of PET dusts and chunks that are shipped to the by-product valorisation plant. In the latter, they are processed or upsized into valuable applications, which is very consistent with the circular economy concept that encourages zero-value waste processes, finishing the cycle of waste to landfill[9] by adding value to all the streams of the recycling process (Webster 2015; Scheel 2016).

---

[7] Largest stadium in Latin America, with a capacity to seat 87,000 people.

[8] Free of volatiles that could affect health or flavor.

[9] See the concept: Circular economy—an industrial system that is restorative by design on: Towards the Circular Economy. Ellen MacArthur Foundation, 2013.

Fig. 10.2 How to dispose of plastic bottles properly programme

10.4.1.4 Customers (Market)

The food-grade recycled PET resin is billed to our shareholders who are also our customers and is shipped in hopper trucks to their bottle manufacturing suppliers and stored in their silos. This is followed by the feeding operation to the production lines by blending it at different percentages with virgin resin to finally produce new bottles with recycled content. The recycled content varies from 25%, 35% or 50% in clear bottles and up to a 100% in the case of most green bottles.

The manufacturing process of the new bottles takes place in two stages; the first occurs in the injection plants, where a preform (like a test tube) is produced and transported to the bottler. The second stage is carried out at the bottler's, where it is blown to the shape of the bottle, then filled up, labelled and packed to be distributed in the market.

Afterwards the consumer buys the bottled products in the market place, consumes them and as owner of the packaging, has the choice of putting the bottle into the "virtuous circle" of recycling through the four steps shown in Fig. 10.2: (1) empty, (2) flatten or crush, (3) re-cap and (4) deposit. After being deposited, the recyclable material enters the inclusive collection stage described in Sect. 10.4.1, keeping the flow of material circling indefinitely.

10.4.1.5 Innovation and Continuous Improvement

The engine for the PetStar innovation and continuous improvement model is located at the centre of the graphic element shown in Fig. 10.1. As seen before, there are four main elements within the model, signifying that the entire organisation permanently seeks to improve the processes to generate value systemically. Innovation is promoted by the collaboration of all the personnel in a programme called "Ideas in Action" (Ideas in Action). Through this programme anyone can register improvement ideas, which, according to their feasibility, can be implemented. This is an awarding process that implies that the owner of the implemented idea receives either a diploma or an economic recognition; the "prize" depends on the idea's impact level on the company's objectives.

10.4.1.6 The Circular Economy Framework of PetStar

PetStar is using some of the circular economy EMF, ReSOLVE[10] framework (Webster 2015; Lacy and Rutqvist 2015), "which offers organisations a tool for identifying circular strategies and growth initiatives", as well as other concepts of circularity applied on other countries (Mathews and Tan 2011). The company uses this framework to leverage the dynamics of the PSBM to take advantage of some of its alternatives—regenerate, share, optimise, loop, virtualisation and exchange—for the current recycling of plastic packaging and link it to the PetStar's regular activities. For instance:

Regenerate Restore materials.

Using transformation processes to regenerate the properties of the PET bottle that has been collected and processed to a food-grade standard and to insert it into a permanent usage system.

PetStar recovers 80,000 tons of post-consumer PET bottles per year, which is equivalent to 70% of the PET its shareholders put on the market and turns them into 50,000 tons of high-quality PET food-grade recycled resin that is reincorporated into new bottles of the Mexican Coca-Cola Industry, the other 30,000 tons corresponding to by-products and small losses of humidity and dirt.

Optimise Prolonging product's use period through closing "almost permanently" the recycling cycle.

The efficient flow of materials through the four main activities of the model (inclusive collection, collection processes, recycling processes and customers/market) constitutes PetStar's circularising process. With these four activities, the value of the PET bottle is maintained for the longest time possible (ideally, in a permanent way) by reincorporating it into the production of new bottles with recycling content.

In 2016, the PET recycled content rate of PetStar's shareholders was 32%.

Optimise How to achieve a zero-plastic leakage level.

Concerned by the challenge of plastic trash that ends up in the ocean and aligned with the initiative called "Operation Clean Sweep" (ACC and PLASTICS 2017), the PSBM seeks to demonstrate that it is possible to avoid leakage of plastic material along the entire chain (bottles, caps, labels, flakes, dust and pellets) due to its full integration, from the recollection of the bottles all the way to the production of the food-grade recycled PET resin.

[10] Towards the circular economy, Ellen MacArthur Foundation, 2013.

Optimise How to get (and share back) most of the scavenger's network.

The living conditions of the close to 24,000 scavengers around the country[11] who supply the feedstock represent a challenging task to dignify this activity. What's more, to optimise their quality of life and efficiency, PetStar is making several efforts to provide them with certainty through fair price, service on time and punctual payment. Also, PetStar promotes the living, health and education conditions of these people, as well as of their families, especially their small children, translating this into better communities, where they are free to decide where to sell the material collected.

Optimise Use renewable energy in its processes.

The plant's cogeneration energy cycle, as well as several systems installed to generate energy under a clean energy cycle, such as wind power, makes the plant a sample of how to reduce energy in the operation.

Loop Keeping materials and products in an iterative cycle of incremental value sharing.

PetStar, by means of several incentives, organisational initiatives and programmes with the scavengers and collectors who are part of the network of suppliers for the collection plants, is working on the process of closing the loop by collecting 80,000 tons of bottles, which represent 10.7% of all new bottles sent to the market (ECOCE 2017). At the recycling plant, it produces 50,000 tons of food-grade recycled PET resin that is permanently circling in a closed loop of new bottles.

The by-products are also processed in other streams of recycled materials of added value that enter other applications, such as a polyester-based polyurethane system for the insulation of commercial refrigerators used by the PetStar shareholders in the point of sale, or the polyolefin caps and labels that are processed into a compound that is used as material in new crates for the refillable bottles.

Exchange Using alternative materials inputs.

A very important element to boost the circular economy is the adequate design of the products. In this respect, PetStar installed a botelloteca[12] and laboratory[13] to promote full compliance of the PET bottles sent to the market with the guidelines for recyclability defined by the "APR Design® Guide for Plastics Recyclability" (APR 2016) to maximise the value of the material when it is recycled (up cycling) (McDonough and Braungart 2002).

[11] Data obtained from PetStar Internal Report (2016).

[12] Showroom with samples of all the bottles available in the market.

[13] Equipped to perform the APR test protocols.

10.4.2 The Zero-Plastic Leakage Framework

The zero-plastic leakage framework and the challenge of avoiding plastic trash in the soil and oceans in alignment with the initiative "Operation Clean Sweep" (ACC and PLASTICS 2017) have been embedded in the core of the PSBM to avoid leakage of plastic material (bottles, caps, labels, flakes, dust and pellets) along the entire value chain system.

10.4.3 The Sustainability Framework

Sustainability is the core of the PSBM, so to guarantee its long-term viability, all our processes are aligned and focused on maximising *social, environmental and economic* values, contributing from the private sector for Mexico to accomplish the 17 objectives of the Sustainable Development Goals (United Nations 2015).

PetStar operation shares the vision of the MCCI to recover the post-consumer bottles from the environment and recycle them efficiently to convert them once again into new bottles with recycled content, thus internalising 100% the social, environmental and economic values in Mexico by contributing locally to a global challenge.

10.4.4 Social Value

The social value of PetStar is measured in terms of employment indicators: almost 1000 direct employees and 24,000 indirect jobs represented by scavengers and collectors in the country. Also, considered are the educational programmes implemented within the communities.

The IDCC project developed in alliance with other CSOs in the municipality of Chimalhuacan, state of Mexico, is a clear example of this social effort.

PetStar's main strength is its human talent, for which internal volunteer programmes have been implemented so that the time and capabilities of all its collaborators can contribute along different specific journeys in the framework of (1) the Global Recycling Day, (2) the Day of the Tree, (3) the Earth Day, (4) the International Day for the Environment and (5) the Day Without Cars. Even further, there are other volunteer programmes to accompany the IDCC children in the celebration of Christmas, which turned out to be a very touching event, organised by the CSO that operates the food service.

The concept of shared responsibility between the society, the public and the private sectors is the only way to face the big challenge of the management and exploitation of the solid residues in Mexico, in which everyone should take its responsibility and do their part (Porter and Kramer 2011).

As part of this shared responsibility scheme, *society as consumers*, when someone buys a bottled product, that person becomes the owner of the purchased product and its packaging. In the case of PET, after a bottled product is consumed, the owner of the bottle has not only the possibility but the responsibility to initiate the process of recycling in the right way, which PetStar promotes with four steps: (1) empty, (2) flatten, (3) re-cap and (4) dispose of the bottle properly.

In fact, Fig. 10.2 shows a scheme of simple actions that decrease 75% of the packaging volume, making the whole recovery chain more efficient, efficiency that can be translated into an important reduction of the carbon footprint generated by the recyclable residue, as well as the caps that are captured and recycled. In addition, when a citizen performs the four-step scheme in a conscious way, he/she is recognising that the bottle is not trash but a valuable recyclable residue. Furthermore, this attitude shows eagerness to participate to guarantee its circularity.

10.4.5 Environmental Value

PetStar's challenging environmental goals for 2020 include (1) neutralising its carbon footprint, (2) neutralising its water footprint, (3) zero-plastic leakage and (4) zero waste. There has been a great evolution towards these goals, since in 2016 we implemented a cogeneration project that decreased the energy consumption in the recycling plant by 15%, plus the use of wind power.

10.4.6 Economic Value

Independently of the positive social and environmental impacts for its shareholders, the PSBM generates value to the supply chain of the Coca-Cola bottlers in Mexico; PetStar guarantees its economic viability by maintaining the price competitiveness of its PET recycled resin versus the virgin resin (as it is substituted) under any market condition. Furthermore, it offers a reasonable financial performance to the committed investment.

To keep a sustainable model under economic feasibility, a circular economy business model (Lewandowski 2016) is achieved through innovation, efficiency and constant improvement in every process, which maintains an effective circularity and, at the same time, obtains positive economic indicators, aligned to the circularity principles and avoidance of plastic leakage.

10.4.7 Philosophy of Excellence

As seen in Fig. 10.1, the PSBM pursues the exercise of its leadership in a very responsible way by promoting sustainable development and the formalisation of the plastic recycling sector in Mexico. In this regard, PetStar seeks to be a reference of excellence at the international level by demonstrating that this activity can be performed with high standards endorsed by its philosophy of excellence (Table 10.2).

Table 10.2 PetStar's philosophy of excellence credentials

Symbol	Name	Description
	Sustainable Development Goals (SDG)	From the private sector, PetStar seeks to contribute with international commitments that Mexico, as a country, has signed up on. One of them is the commitment to accomplish the 17 Sustainable Development Goals (SDG) of the 2030 Agenda for Sustainable Development (United Nations 2015). Most of the processes included in PetStar's business model point to those targets, as indicated previously
	Earth Charter Initiative	In 2014 PetStar adhered to the Earth Charter Initiative that is an ethical framework to build a just, sustainable, and peaceful global society for the twenty-first century. It seeks to inspire in all people a new sense of global interdependence and shared responsibility for the well-being of all humanity, the greater community of life and future generations. It is a vision of hope and a call for action (The Earth Charter 2008)
	The Global Compact Mexico	In 2016 PetStar adhered to the UN Global Compact in Mexico, which is the world's largest corporate sustainability initiative and a call to companies to align strategies and operations with universal principles on human rights, labour, environment and anti-corruption and take actions that advance societal goals as well (United Nations Global Compact 2004)
	Socially Responsible Company (ESR)	Every year since 2015 PetStar received the "Socially Responsible Company" (CSR in its English acronym) acknowledgement granted by CEMEFI (Centro Mexicano para la Filantropia) as well as its affiliation to the "ESR del 1% de inversion social" (Socially Responsible Company of the 1% of social investment) meaning that at least 1% of profit before taxes will be donated to any social cause and contribute to a more just society and a better country, and starting in 2016 it engaged in the programme "ESR Comprometidas en favor de la inclusion social" (Socially Responsible Companies committed in favour of social inclusion) (CEMEFI 1994)
	Top Companies	Every year since 2015 PetStar measures its organisational culture through Top Companies, which is an external Mexican consultant who shares the results with Grupo Editorial Expansion. This publishes the ranking as Super Empresas Expansion (Expansión 2016), which refers to "the place where everybody wants to work", achieving in 2015 the 44th place of companies in the category of 500 to 3000 employees and in 2016 climbed to the 41st place

(continued)

Table 10.2 (continued)

Symbol	Name	Description
	Clean Transportation	For the good environmental performance of its fleet of trucks in 2016 PetStar achieved registration in the Clean Transport Programme granted by SEMARNAT (Secretaría del Medio Ambiente y Recursos Naturales) (SEMARNAT n.d.)
	Operation Clean Sweep (OCS)	Since 2015 PetStar embraced the "Operation Clean Sweep" (OCS) initiative (ACC and PLASTICS 2017), an international programme designed to prevent resin pellet, flake, and powder loss to help keep this material out of the marine environment.
	LEED Platinum	In 2015, the PetStar Museum Auditorium located in the recycling plant in Toluca achieved the LEED (Leadership in Energy and Environmental Design) Platinum certification awarded by the US Green Building Council as a sustainable building, becoming the first museum in Latin America to achieve this level of certification
	Clean Industry	Every year since 2015 the recycling plant in Toluca achieved the distinction of Clean Industry at a federal level granted by PROFEPA (Procuraduría federal de Protección al Ambiente), as well as clean industry at the state level granted by PROPAEM (Procuraduría de Protección al Ambiente del Estado de México)
	PASST	In 2016 PetStar signed an agreement with STPS (Secretaria del Trabajo y Previsión Social) to incorporate voluntarily all its plants into "Self-Management Programme in Health and Safety at Work". In 2016, the collection plants in San Luis Potosi, Merida, Monterrey, Acapulco and Guadalajara were certified and in 2017 the collection plants in Toluca, Ecatepec and Queretaro, as well as the by-product valorisation plant and the recycling plant, both in Toluca, will be certified
	ISO 9001	In 2014, the recycling plant in Toluca achieved the ISO 9001 certification, then in 2015 the collection plant in Ecatepec and finally in 2016 all other seven collection plants, as well as the by-product valorisation plant
	ISO 14001	In 2015, the recycling plant in Toluca obtained the ISO 14001 certification, and in 2017 the eight collection plants and the by-product valorisation plant will be certified

(continued)

Table 10.2 (continued)

Symbol	Name	Description
	ISO 22000	In 2015, the recycling plant in Toluca achieved the ISO 22000 certification. This is very challenging for a recycling plant
	ISO 50001	In 2016, the recycling plant in Toluca achieved the ISO 50001 certification, and in 2017 the eight collection plants and the by-product valorisation plant will be certified
	OHSAS 18001	In 2014, the recycling plant in Toluca achieved the Occupational Health And Safety Standard (OHSAS 18001) certification

10.4.8 PetStar's Mission, Vision and Values

Based on the above framework, the company has stated its mission, vision and values.

The mission has been oriented to promote the preservation of the environment in benefit of future generations by implementing innovative solutions that make the PET bottle sustainable, with efficient collection and recycling to produce high-quality PET food-grade recycled resin with social responsibility.

The company's vision is to become a worldwide reference of excellence through the inclusive recycling of the PET bottle that with talent, commitment, innovation and well-being adds value with social and environmental responsibility to its clients, thus contributing to the solution of climate change.

Values, such as *safety*, *honesty*, *responsibility*, *service*, *reliability* and *respect*, are internalised in PetStar by living them every day, contributing to global sustainability and respect for the Earth Charter's (The Earth Charter 2008) fundamental principles.

10.5 The Sustainable Benefits of PetStar Circular Operation

Sustainability is the core of the PetStar sustainable business model, so to guarantee its long-term viability, all company processes are aligned and focused on maximising *social, environmental and economic* values, contributing from the private sector to Mexico to accomplish most of the 17 objectives of the Sustainable Development Goals (United Nations 2015).

PetStar operation shares the vision of the Mexican Coca-Cola industry to recover post-consumer bottles from the environment and recycle them efficiently to convert them once again into new bottles with recycled content, internalising the social, environmental and economic values 100% in Mexico and contributing locally to a global challenge.

In summary, PetStar has implemented a well-planned corporate strategy that supports pursuing the following objectives:

1. *Social value*
   (a) Dignifying the labour of scavengers
   (b) Optimising scavengers' living conditions
   (c) Providing fair and stable income to scavengers and collectors
   (d) Giving access to education, food and medical services to the children of scavenger communities
   (e) Promoting the creation of public policies for integrated waste management with the inclusion of scavengers
   (f) Promoting well-being for almost 1000 direct employees
   (g) Creating internal volunteer programmes
   (h) Using the museum auditorium to promote environmental awareness in society
2. *Environmental value*
   (a) Recuperate the bottles to prevent pollution of soil and water, as in 2016, when almost 80,000 tons, equivalent to 70% of the bottles that its shareholders sent to the market, were collected.
   (b) Facilitate the sustainability of shareholders' PET packaging, as in 2016, when an average of 32% of PET recycled content was used in their bottles and in most of their green bottles, 100%.
   (c) Drastic CO_2 emission mitigation versus virgin resin (towards neutral carbon footprint for 2020).
   (d) Prevention of plastic leakage of bottles, caps, labels, flakes, dust and pellets in the system (towards zero-plastic leakage for 2020).
   (e) Zero-value waste transformation to valuable assets (towards zero-waste for 2020).
   (f) Use of renewable energy sources.
   (g) Efficient use of water in the process (towards neutral water footprint for 2020).
   (h) Promote the PET Mexican packaging industry to comply with the "APR Design® Guide for Plastics Recyclability" to improve the recycling viability of the bottles in the market.

3. *Economic value*
   (a) Guarantee to its shareholders a price for the food-grade recycled PET resin equal or better than virgin under any market condition (such as oil price fluctuation and a balance of supply–demand recycling model).
   (b) Ability to produce 50,000 tons per year of food-grade recycled PET resin with an installed capacity of 40,000 tons per year.
   (c) Profitable operation, as it generated the following EBITDA: 5.6% in 2014, 11.2% in 2015 and 15.6% in 2016.

10.6 Conclusions

> The recycling industry exhibits the win-win-win properties of sustainable growth, generating employment social equity and being enviro-friendly; but the trade-offs are considerable in combining these aims. (UNIDO 2015)

Solid waste management in developing countries today is threatened by several negative externalities, the society's poor sustainable culture, the deficient infrastructures, the informal economy where the scavengers operate, their low levels of human development, non-operable or lack of public policies, the astronomical volume of residues generated and unmanaged by the public services, the huge loss of value by burying the residues in landfills and the inability to contain the leakage to natural ecosystems, contaminating soil and water.

The factors that create a challenging reality to the plastic packaging recycling industry, in addition to its own threat of lower oil price and the light weight of the material, make this industry a great focus of interest for entrepreneurs, environmentalists and innovators, looking for better ways to attack such negative issues and convert these externalities affecting a low-profile industry into a sustainable business.

One of the innovation practices currently used to attack the above mentioned threats is a correct implementation of circular techniques. The term "waste is food" coined by McDonough and Braungart (2002) summarises the circular approach, and in industries like plastic packing, this concept acquires significant meaning, given the large volume of residues that today pollute the planet.

PetStar has been able to close the loop by optimising resources and applying non-usual activities that link the scavengers (suppliers of feedstock) to the final product in a win-win process. The company has developed a programme of inclusion of the scavenger community for better education, health and living conditions. In the long term, this activity will reward the whole society with social, environmental and economic benefits.

Design for circularity is a key principle of the circular economy approach, and for this purpose PetStar has developed an infrastructure to analyse the technical specifications of the available bottles and all their components, like material, labels, caps, glues, barriers, etc., compare them to the APR Design® Guide for Plastics

Recyclability and offer alternatives to the bottle manufacturers and marketing departments to make sure that all the bottles available in the market are in full compliance, maximising their potential value when they are recycled.

With the implementation of the PetStar sustainable business model, the company wants to demonstrate that recycling is the path to the sustainability of plastic packaging, specifically for the PET bottle in a closed loop bottle-to-bottle cycle. By doing this with very high standards in processes and management, under the framework of its philosophy of excellence, the company wants to break with the paradigm that recycling is a dirty, low-quality, high-risk activity. On the contrary, PetStar is a very clean, high-quality, safe process with world-class certifications and recognition, translating into a profitable business that generates social, environmental and economic values.

Another important conclusion is PetStar's ability to optimise the number of cycles a bottle can have to close the loop and minimise the use of petrochemical components, as one of the alternatives to decoupling plastics from fossil feedstock, maximising the environmental value of the material.

The main lesson learned from this chapter is how to run a company that can close the sustainability cycle, upgrade social inclusion and achieve a zero-waste to landfill approach by optimising the recovery of PET bottles and, at the same time, be economically feasible and competitive.

References

ACC & PLASTICS (2017) Operation clean sweep. https://opcleansweep.org/Standalone-Content/Operation-Clean-Sweep-Manual-PDF-Version.pdf/. Accessed 29 Apr 2017

APR (2016) The APR design guide for plastic recyclability. http://www.plasticsrecycling.org/images/pdf/design-guide/Full_APR_Design_Guide.pdf. Accessed 29 Apr 2017

CEMEFI (1994) Empresa Socialmente Responsable. CEMEFI. https://www.cemefi.org/esr/. Accessed 29 Apr 2017

ECOCE (2017) ECOCE 15 AÑOS Informe 2017. http://ecoce.mx/files/Informe-ECOCE-2017.pdf. Accessed 5 May 2017

EMF (2013) Towards circular economy. Ellen MacArthur Foundation Publishing, London

EMF (2016) The new plastics economy—Rethinking the future of plastics. Ellen MacArthur Foundation Publishing, London

EMF (2017) The new plastics economy—catalysing action. Ellen MacArthur Foundation, London

Expansión (2016) Ranking Expansión Súper Empresas. http://expansion.mx/rankings/2016/05/20/super-empresas-expansion-2016. Accessed 29 Apr 2017

IRR (2011). Regional inclusive recycling initiative. http://inclusiverecycling.com/. Accessed 29 Apr 2017

ISO (2005) ISO 22000 food safety management. https://www.iso.org/obp/ui#iso:std:iso:22000:ed-1:v1:es. Accessed 5 May 2017

ISO (2011) ISO 50001 energy management. https://www.iso.org/obp/ui#iso:std:iso:50001:ed-1:v1:en. Accessed 5 May 2017

ISO (2015a) ISO 9001 for Quality Management System. https://www.iso.org/iso-9001-quality-management.html. Accessed 5 May 2017

ISO (2015b) ISO 14001 for environmental management. https://www.iso.org/standard/60857.html. Accessed 5 May 2017

Lacy P, Rutqvist J (2015) Waste to wealth: the circular economy advantage. Palgrave Macmillan, New York
Lewandowski M (2016) Designing the business models for circular economy—towards the conceptual framework. Sustainability 8(1):43. https://doi.org/10.3390/su8010043
Mathews JA, Tan H (2011) Progress toward a circular economy in China. J Ind Ecol 15(3):435–457. https://doi.org/10.1111/j.1530-9290.2011.00332.x
McDonough W, Braungart M (2002) Cradle to cradle: remaking the way we make things. North Point Press, New York
McGeough H (2013). Worlds largest PET food grade recyclers. PCI PET Packaging Resin and Recycling, Ltd
Medina M, Smith C (2013) Social inclusion in Mexico's PET plastic recycling. United Nations University. Accessed 5 May 2017
OHSAS (2007) Occupational health and safety standard. https://www.bsigroup.com/es-MX/bsohsas18001-salud-seguridad-ocupacional/. Accessed 5 May 2017
PetStar (2016) Assessment of PetStar's sustainability performance (Life Cycle Assessment). PCI WoodMackenzie
Porter ME, Kramer MR (2011) Creating shared value. Harv Bus Rev 89(1):62–77
PROFEPA (1992) Industria Limpia Federal. http://www.profepa.gob.mx/innovaportal/v/26/1/mx/programa_nacional_de_auditoria_ambiental.html. Accessed 5 May 2017
PROPAEM (2014) Industria Limpia Estatal. http://propaem.edomex.gob.mx/certificacion_industrial. Accessed 5 May 2017
Scheel C (2016) Beyond sustainability. Transforming industrial zero-valued residues into increasing economic returns. J Clean Prod 131:376–386. https://doi.org/10.1016/j.jclepro.2016.05.018
SEMARNAT (2006) Bases para Legislar Prevencion de la Gestion Integral de Residuos. http://siscop.inecc.gob.mx/descargas/publicaciones/bases_legislar_prevencion_gestion_integral_residuos.pdf. Accessed May 10 2017
SEMARNAT (2017) Programa de Transporte Limpio. Secretaria del Medio Ambiente y Recursos Naturales. http://www.gob.mx/semarnat/acciones-y-programas/programa-de-transporte-limpio-26305. Accessed 5 May 2017
STPS (2008) Programa de Autogestión en Seguridad y Salud en el Trabajo (PASST), Mexico. http://autogestion.stps.gob.mx:8162/pdf/LineamientosGenerales2008.pdf. Accessed 5 May 2017
The Earth Charter (2008) Earth Charter initiative handbook. http://earthcharter.org/invent/images/uploads/Handbook.ENG.pdf. Accessed 5 May 2017
Top Companies (2016) Top companies ranking. http://thetopcompanies.com/. Accessed 5 May 2017
U.S. Green Building Council (2017) Leadership in energy & environmental design certification. http://www.usgbc.org/LEED/. Accessed 5 May 2017
United Nations (2015) Sustainable development knowledge platform. https://sustainabledevelopment.un.org/?menu=1300. Accessed 29 April 2017
United Nations Global Compact (2004) The ten principles of the UN Global Compact. https://www.unglobalcompact.org/what-is-gc/mission/principles. Accessed 29 April 2017
Webster K (2015) The circular economy: a wealth of flows. Ellen MacArthur Foundation Publishing, London

Chapter 11
Social and Environmental Life Cycle Assessment (SELCA) Method for Sustainability Analysis: The Jeans Global Value Chain as a Showcase

María-Laura Franco-García, Willem Haanstra, Marten Toxopeus, and Boelo Schuur

Abstract In this chapter the concepts of social life cycle assessment and combined social and environmental LCA were explored through the application of existing LCA methods to the global value chain of jeans. The social and environmental life cycle assessment (SELCA) method resulted from this explorative research that aims to contribute to the battery of impact assessment tools of products whose value chain scope is multinational (global). From a broader perspective, SELCA has a double-folded purpose to (i) identify opportunities for environmental and social improvement at any of the value chain phases of products, for remediation goals, and (ii) predict the environmental and social performance of different ways (scenarios) to produce the same product, using it as a product design tool. To simplify SELCA development, it was decided to use a single product (jeans) as a showcase from the global textile sector. In this showcase, four scenarios for jeans assembly were compared; three of them were defined under the circular economy principles by including recycled materials (cotton, PET and nylon 6) during the yarn production. During the application of the SELCA method, some new challenges were encountered related to inventory analysis, in particular during data acquisition for social inventories. This is later mainly due to the extensive list of key stakeholders

M.-L. Franco-García (✉)
University of Twente, Enschede, The Netherlands
e-mail: m.l.francogarcia@utwente.nl

W. Haanstra · M. Toxopeus
Faculty of Engineering Technology, Design, Production and Management, University of Twente, Enschede, The Netherlands
e-mail: w.haanstra@utwente.nl; m.e.toxopeus@utwente.nl

B. Schuur
Faculty of Science and Technology, Sustainable Process Technology, University of Twente, Enschede, The Netherlands
e-mail: b.schuur@utwente.nl

M.-L. Franco-García et al. (eds.), *Towards Zero Waste*, Greening of Industry Networks Studies 6, https://doi.org/10.1007/978-3-319-92931-6_11

for the showcase and the qualitative nature of social metrics. This list starts with cotton cultivators from different countries where regulations and codes of conduct seem to have contextualised interpretations and consequently different levels of implementation. In this regard, governmental intervention to instrument the transition towards suitable social/environmental performance along the global jeans value chain was also discussed in this chapter.

Keywords Combined social and environmental life cycle assessment · SELCA · Jeans · Global value chain · Stakeholders · Circular economy

11.1 Introduction

The attention of scholars to the field of social life cycle assessment (S-LCA) has flourished over the last years (Petti et al. 2016). The development of social impact assessment methods is considered to be challenging, especially due to the qualitative and subjective nature of the social conducts. Yet, combined life cycle assessment is considered to be in its infancy until it can be proven that 'it works' (Jørgensen et al. 2013). This interesting gap in the LCA literature inspired this research which resulted in the social and environmental life cycle assessment (SELCA) method. This method combines the goal and scope definition although the social and environmental impacts are differently assessed. Hence SELCA was initially developed to challenge Jorgensen's claim to find out if 'it works'. To do this we used the case study on the global (and partially circular) value chain of denim jeans. Due to the environmental and social controversially textile sector, a pair of jeans were used as a showcase to simplify the social LCA development and the comparison of some of the circular and non-circular jeans scenarios.

The denim jeans were used as a showcase, and four (baseline and recycling) scenarios for the yarn production were assessed: 100% virgin cotton (VC), mix of VC with recycled cotton, mix of VC with recycled PVC and mix of VC with recycled nylon. All scenarios were assessed through the SELCA method. The jeans' value chain goes beyond local conditions with suppliers and production phases outside the consumers' market. Even further, textile (jeans) value chain is the subject of a wide variety of identified actors throughout its life cycle in a global level. Literature already pointed out that in a combined LCA, alignment of the goal, scope definition and the functional unit can be seen as a challenging matter (Macombe et al. 2013). Also, the global value chain raises the issue of the identification of the most important actors and ways of their inclusion in the assessment. The recycling scenarios are evaluated by using the profiling results of the environmental and social dimensions.

The research question driving this work was formulated as the following: *How can social aspects be integrated into (traditional) environmental life cycle assessment by using denim jeans as showcase*? This research was focused to answer this question by using a showcase for the textile sector: jeans. The information gathered and analysed for such purpose is presented in this chapter which started with the literature revision. This was focused on the life cycle assessment and its challenges (Sects. 11.2 and 11.3), and some case studies were explained to illustrate LCA implementation (Sect. 11.4). Based on the state of art of literature, choices were made with regard to the development of the SELCA method and were highlighted in Sect. 11.5. The proposed methodology is explained by means of the application of the jeans case study (Sect. 11.6). Lastly, conclusions and recommendations are presented in Sect. 11.7.

In order to centre the attention on SELCA development, the nexus between LCA and circular economy (CE) does not make part of this particular chapter, but further elaboration on this nexus is referred to Chap. 13.

11.2 The Social LCA and Combined LCA (the Concepts)

In the late 1990s, S-LCA started gaining attention. In 1996 O'Brien (1996) proposed a methodology to integrate the social and environmental dimension, called the 'Social and Environmental Life Cycle Assessment'. They stated that an integrated assessment provided a more complete potential impact assessment of a life cycle. This can be referred to as life cycle thinking. UNEP states: 'Life Cycle Thinking is about going beyond the traditional focus on production sites and manufacturing processes, so that the environmental, social and economic impact of a product over its whole life cycle are also included' (UNEP 2009). This challenges the conventional resources management and pollution prevention mind-set that is present at production sites.

Over the last few years, many paradigms arose that aim to improve the sustainable performance of products and services, socially, environmentally and economically, defined as the green economy (Barbier 2012). UNEP provided a list of practical concepts and approaches of green economy that includes resource efficiency, cleaner production, the waste hierarchy, circular economy, LCA and CBA[1] (UNEP 2011). Some of these concepts are used in the front end of product development, such as circular economy. This concept focusses on improving effective resource use of products and services. For further description of circular economy, see the introduction chapter of this book. Seeing other concepts, such as LCA, are evaluative tools used at the back end, life cycle tools are most commonly used to assess the sustainability of an applied concept, like circular economy.

The first studies of environmental life cycle assessment date back from the 1970s, the period in which environmental issues started drawing public attention (Guinée 2016). Life cycle assessment is defined as a technique that is used to quantify environmental impacts of a product or service over its life cycle, including raw material extraction, manufacturing, distribution, use and disposal (ISO 2006). During the first decade of the twenty-first century, new approaches were developed,

[1] CBA stands for cost-benefit analysis.

drawn on the triple bottom-line approach, and the social LCA (SLCA) and life cycle costing (LCC) methods were proposed. The integrated approach of life cycle thinking with the triple bottom line (Elkington 1994) is referred to as Life Cycle Sustainability Assessment (LCSA) (UNEP 2011). Kloepffer (2008) defines Life Cycle Sustainability Assessment as the formula LCSA = LCA + LCC + SLCA. LCA stands for the ISO 14040 environmental assessment, LCC is a type of assessment focused on the economic dimension and SLCA represents the social dimension. Guinee builds on the definition of LCSA by adding two dimensions: firstly, broadening the level of analysis from product level to sector level and economic level and, secondly, the deepening of the LCA, from physical relations to economic and behavioural relations.

By combining the triple bottom line in an LCA, a truly holistic representation of sustainability of products can be assessed (Kloepffer 2008). The original concept of LCA only dealt with the environmental impact, as defined in ISO 14040, whereas nowadays it rather relates to the concept of LCSA. SLCA is in line with LCA but adapted for the social aspects. Therefore SLCA also consists of the four phases: (i) goal and scope definition, (ii) inventory analysis, (iii) impact assessment and (iv) evaluation (Weidema 2005). Social life cycle assessment is a methodology that aims at assessing the potential social and socio-economic impact, both positive and negative, of products and services from a life cycle perspective (UNEP 2009). The SLCA method uses information from company, plant and process levels for the whole life cycle of a product. SLCA has similar applications to traditional LCA, such as management, labelling and assessment of alternatives (Chhipi-Shrestha et al. 2015). However, in SLCA the ultimate goal is the *well-being of stakeholders* over a product's life cycle (UNEP 2009).

After increasing attention on social assessment, many approaches towards SLCA have been developed (Hunkeler 2006; Weidema 2006; Norris 2006; Dreyer et al. 2006). Jørgensen et al. (2008) and Parent et al. (2010) found that there was a large amount of diversity among existing SLCA approaches. For that reason the 'United Nations Environmental Program' identified the need for the establishment of an international aligned framework to assess social impacts across product life cycles. This resulted in the SLCA guidelines, consisting of 5 stakeholder categories (workers, local community, society, consumers and value chain actors) and 6 impact categories with 31 subcategories, which are characterised using more than 100 inventory indicators (UNEP 2009). The guidelines were followed by the methodological sheets which provide practical guidance for conducting SLCA case studies (Benoît-Norris et al. 2011).

After publication the UNEP guidelines have been widely adopted in the development of social life cycle impact assessment methods (SLCIA) up until now (Chhipi-Shrestha et al. 2015). According to Chhipi-Shrestha et al. (2015), many different impact assessment methodologies have been used due to the lack of specific SLCIA methodology. Mattioda et al. (2015) found that the lack of common approach is due to the *lack of shared references* in the field of SLCA. This brings us to the statement of Jørgensen et al. (2013) saying that SLCA is still in its infancy, since a mature discipline is characterised by a large amount of shared references.

With the purpose to find the differences in approach Parent et al. (2010), compare three different SLCIA methods which all claim to use the UNEP guidelines in their method and their results. It was found that two main impact assessment method (IAM) categories can be observed. Type 1 is the *performance reference point method*, which uses additional information, like international standards. Type 2 is the *impact pathway method*, which translates the inventory indicator into a midpoint and endpoint indicator. With type 1 the outcome will be a social performance along the product life cycle. By use of type 2, one can assess social impact quantitatively related to the functional unit, however for a smaller set of social issues. Chhipi-Shrestha et al. (2015) subdivide these IAM categories. The performance reference point method is of qualitative nature and can distinguish the checklist methods, scoring methods and the social hotspot database method. The impact pathway methods are of quantitative nature, where one can distinguish the empirical method and the environmental LCI database method.

Within this study the UNEP guidelines were used, given the completeness and the wide use of these guidelines among scholars. The scoring method is applied for the assessment of social impact, because it provides the possibility of assessing social conduct without the need for statistical data, which suits the aim of applying the method for a broad, general and indicative purpose.

11.3 Challenges in SLCA and Combined LCA (Developments)

The combined social and environmental LCA has been studied by various scholars who identified multiple challenges during its conceptualisation, development and implementation phases. Those previous works have shaped, to some extent, the approach used in Sect. 11.5 to adapt the LCA method to global value industrial sectors, i.e. the textile case explained in Sect. 11.6.

The social perspective of LCA expands the traditional physical limitations of the product manufacturing to other contextual regions from which resources are extracted and regions where people play a vital role from inside and outside of the manufacturing facilities. Even further, some research has already explored (Sutherland J.W et al. 2016) the effect manufacturing has on society, as a whole. It was found that people have two roles in the value chain. People's needs generate demand that is met by products and services. On the other hand, manufacturing products and services generate employment, which ensures people to be able to sustain their lives. The social conditions behind the life cycle of a product are invisible to the consumers as well as they might be to the producer. Producers often don't even have insight in the practises of their suppliers earlier in the value chain (Dreyer et al. 2010). Though, stakeholders now demand accountability of the company for the behaviour of value chain actors, over which the company obviously has limited control. Since there is more and more pressure on brands to know the social impact of their products, they have to be able to acquire knowledge on the social conduct of the other actors in their value chain.

In addition to this trend, companies become more and more aware of the increasing attention the field of sustainability receives by consumers as well as by competitors. According to Sutherland J.W et al. (2016), there is a high value in integrating the three different dimensions of sustainability in one assessment method. The multiple dimensions provide insight in the interrelatedness of systems. Economic activities, for example, are inevitably related to social and environmental impacts (Sutherland J.W et al. 2016). On the other hand, preserving ecology might go at the expenses of communities. Hence, it can be safe to say that all activities have an economic and social side in addition to the environmental dimension.

Recognising and avoiding trade-offs between the dimensions of a product life cycle may be seen as the goal of the combined LCA. The scope of this study considers the combination of the social and environmental dimension into one LCA, which brings some implications and challenges. As priory mentioned, the level of maturity of the social LCA is very low comparable to the level of the environmental LCA. Therefore this literature review also touches upon the main immaturity issues of the SLCA.

Before the main challenges in literature are identified, it must be noted that the combination of the triple bottom line within one LCA tool is in an early stage of development. The formula for Life Cycle Sustainability Assessment, as defined by Kloepffer (2008), is kept mostly as a concept and does not receive much attention to be further developed or applied. Finkbeiner et al. (2010) do elaborate on this theoretical development by employing the Life Cycle Sustainability Triangle (LCST) and the Life Cycle Sustainability Dashboard (LCSB). However, these two methods try to address the challenge of making LCSA results understandable for its target audience of decision makers and do not focus on actual use of the method. All in all, the three main challenges presented in the following, essentially, are derived from a variety of studies.

First of all, the difference in impact allocation is an obstacle for integration of the environmental and social dimensions into one LCA. It is not yet the intention to discuss about a 'sustainability' LCA which includes also the economic dimension. Impact allocation in the environmental dimension is modelled via scientific supported indicators and pathways. In the social dimension, there is a trade-off between the use of generic data and site-specific data (Dreyer et al. 2006) of quantitative or qualitative nature. Where generic data have an advantage in relation to practicality, only site-specific data offer the proper accuracy according to practitioners (Jørgensen et al. 2008). Some people from companies consider the differences within the product chain as negligible and believe that generic data give a sufficient picture of the associated social impacts, whereas others think each individual company in the product chain has to be assessed, because they all have a different conduct to which the social impact is connected. To add on this, Macombe et al. (2013) find that data and allocation methods are very entity level specific. A company, region and country, all, are interested in different impacts, and therefore all demand a different assessment.

Environmental impacts arise from the nature of the processes, what determines a causal link between the two. Spillemaeckers et al. (2004) and Dreyer et al. (2006) mention that social impact must be seen as something that comes from the conduct of the company (the choices the company makes). Social impact hardly has any relation with the products and processes themselves. The share of the company in the process in the value chain should determine the weight of the conduct that has to be allocated. Often it is even the case that companies cannot control the actions of their foreign suppliers; wherefore Dreyer et al. (2010) in a later paper argued that companies' efforts towards social conduct should be taken into account instead, whereas Kloepffer (2008) and Chhipi-Shrestha et al. (2015) believe the list of social topics must include more issues that relate to products and processes. Schmidt et al. (2004) agree with this focus, because it is used in ELCA[2] as the basis for the assessment too. Most of the indicators apply to countries or regions, and some social indicators even include certain political aspects. The consequence is that indicators of another reference level have to be used and indirectly via methodological assumptions have to be related to the product or process.

Where the ELCA impacts are based on natural sciences, the impact categories of the UNEP guidelines are mainly based upon political consensus, according to Arvidsson et al. (2015). Some social topics defined by these guidelines can be interpreted differently depending on cultural background and on political, ethical and ideological views (Baumann et al. 2013). Non-SLCA social science literature shows that the relationship between social topics and entities is complex. Ambiguousness and complexity of social topics make Arvidsson et al. (2015) question their use in SLCA, what even makes them question the general use of stakeholders and indicators in SLCA. They propose the use of social science sources together with the concept of impact pathways what could lead to scientifically justified topics. In addition, Chhipi-Shrestha et al. (2015) suggest to *combine impact pathways and performance reference points* for further development of the SLCA method.

Secondly, defining the same goal and scope for the SLCA as for the LCA has the advantage of maintaining consistency. Foolmaun and Ramjeeawon (2013) did a study to compare different disposal scenarios of PET bottles using both environmental LCA and SLCA. In this study three different approaches to define the system boundary were explained: (1) narrow the system boundary down to the parts of the life cycle which are directly influenced by the company performing the assessment; (2) include the entire life cycle but exclude processes that do not significantly change the overall conclusions of the study; and (3) only include the organisations that would also be involved in an environmental LCA. When product systems become complex, system boundary inconsistencies between ELCA and SLCA will increase (Wu et al. 2014). For example, transport has to be included in ELCA but cannot be in SLCA. Therefore the challenge is to find a flexible way of defining cut-off criteria (Chhipi-Shrestha et al. 2015). In accordance to Macombe et al. (2013), the system boundary and functional unit can simply not always be the same

[2] In this paper, ELCA corresponds to the LCA focused only on the environmental impact assessment.

among the two dimensions. Also, they claim that the role of the functional unit in a combined ELCA and SLCA is the object of scrutiny.

Thus, there is not yet consensus on cut-off criteria for defining the significant processes, the system boundary and the weighing system (Dreyer et al. 2006). However, in order to make the scope of the inventory analysis workable, prioritisation or cut-off criteria are needed for identifying significant organisations for the side-specific data collection.

The third implication is that the life cycle inventory preferably is the same for the environmental and social LCA (Kloepffer 2008). Yet, what makes this challenging is the difference in nature of societal and environmental data. Data in the social field change faster and therefore require more and regular updates (Wu et al. 2014). Also phenomena like market competition might produce different social impacts from the defined impacts (Dreyer et al. 2006). Whereas environmental LCA strives to generic data which is regionally or globally applicable, the social LCA requires more detail of the contextual resolution of impacts. Additionally, a difficulty in impact assessment in SLCA is the qualitative nature of the indicators as well as the scaling of some of them (knowing what is good and what is bad). On the other hand, quantitative assessment requires generalisation, which is difficult when the impacts resulting from a change differ so much among countries (Macombe et al. 2013). Generic data can be made national, sector or company specific, for example, by the social hotspot database (Hunkeler 2006).The social hotspot database (SHDB) shows value in the case study of Benoit as an innovative tool that offers top-down visualisation of a product supply chain's potential impacts (Benoit-Norris et al. 2012).

To summarise, three main issues for combination of the environmental and social dimension in LCA have been identified. First, the difference in nature of data leads to several implications. The indicators and related impact allocation are for the environmental dimension scientifically justified, whereas for the social dimension, trade-offs have to be made between generic- and side-specific data measured qualitative or quantitatively. Even the justification of the indicators themselves is susceptible to ambiguousness, and measuring them is prone to subjective judgement. Besides, the level of detail of the data and their relation with the product system differ per dimension. Second issue regards the goal and scope definition that has to be equal to both dimensions. Especially the system boundary and the functional unit provide challenges in this. Third and lastly, the process of inventory analysis greatly differs among the two dimensions.

11.4 State of Art of Social and Combined LCA Methods Implementation (Case Studies)

In this section, first, some studies that applied the concept of combined LCA are mentioned. Thereafter the gap in the performed case studies on global value chains is highlighted, followed by the state of the art in indicating social issues in the global textile value chain.

Ciroth and Franze (2011) assessed the social impact for cut roses and compared two types of roses. The assessment was based on the UNEP guidelines via the scoring method. The study was conducted by colour coding for different social impact levels, then weighted and aggregated. These levels were defined on the basis of qualitative data, like internationally accepted minimum performance levels. A different case study on the combined social and environmental LCA was performed by Foolmaun and Ramjeeawon (2013) on the topic of PET bottle disposal alternatives. The study showed that recycling is not only beneficial from an environmental viewpoint but can also contribute to improve social impact.

The case studies above are two examples out of a small collection. For this reason, according to Ramirez P and Petti L (2011), more case studies are needed in order to find out where the SLCA methodology is still weak. Petti et al. (2016) studied the SLCA methodology through its application to case studies and found that there are an increasing number of implemented case studies in the field. It was found that the research object of study is either a product or a service, mostly being in the manufacturing sector (26%) or in the agriculture sector (26%). The study of Wang et al. (2016a) shows that most case studies use country level data, instead of regional or company-related data, to assess the general social impact of the process. According to Petti et al. (2016), the social context of the region (developed or developing country) does not influence the number of case studies. However, 40% of the studies have been conducted in Europe, while in Europe the study contributes less due to the lower level of social concerns. Reason for this could be the higher and easier availability of data. Lack of data is often due to lack of transparency in or no cooperation of companies in the supply chain (Traverso et al. 2016). What is not considered by Petti et al. (2016) is the level of globalisation of the value chain of the product under consideration in the case study. Also the review of case studies in literature of Wang et al. (2016b) does not consider 'level of globalisation of value chain' as a factor.

From this, it was concluded that the global value chain (GVC) is a topic that has received very little attention. However, the GVC is one of the main causes of social issues in product life cycles. Globalisation and liberation of trade have led to the movement of low skill, for instance, an intensive labour of textile value chains in developing countries (Los et al. 2015; OECD 2004). The lowering of prices of textile products makes European retailers put more pressure on suppliers to further reduce costs (Taplin 2006). Suppliers consequently reduce on the aspects which they still have in control such as labour conditions (Taplin 2014).

In addition to the socio-economic impact, the geographical shift in manufacturing also is likely to have an environmental impact (Mair et al. 2016). In the study of Mair et al. (2016), both, environmental and socio-economic, indicators are assessed on impacts of Western European textile and clothing consumption between 1995 and 2009. Whereas literature largely assessed these aspects separately, one of the aims was to find tensions between different sustainability goals. Interesting to note is that the analysis makes use of GVC indicators (Timmer et al. 2013) that assess impacts in the production stages of a product's life cycle.

The review of these case studies in combination with the conclusions drawn from the literature review has helped to construct and support the goals of this study. Besides, based on the challenges and arguments in earlier studies, it was possible to make the needed choices to develop the SELCA methodology.

11.5 Development of SELCA Methodology

The purpose of the suggested methodology is presented firstly here, stressing simultaneously on its relation to current literature. In the paragraphs thereafter, the reasoning behind the construction of the most important parts of the method is given. The constitution of the goal and scope within the SELCA are further discussed in Sect. 11.5.1, while the role of and inconsistencies in the inventory analysis of the SELCA are described in Sect. 11.5.2. To finally the selection of the subcategories and indicators were explained in Sect. 11.5.3, and the development of scorecards associated with social indicators and weighing factors is there given.

The SELCA social impact assessment methodology was primarily developed for initial and broad evaluation of social aspects, such as risk profiles of industries in entire countries. The method needs to allow inclusion of specific inventory when more data become available. Development of the methodology has been done under a set of limited conditions, such as the time constraint of 2 months and the participation of a handful of researchers. For these reasons the methodology was solely meant for explorative use on the GVC textile case study, which takes the attention in Sect. 11.6. The goal of the textile case study was to compile and evaluate the environmental and social consequences of different life cycle scenarios for fulfilling one function. The deployment and application of the methodology aims to add a new perspective to some of the challenges illustrated in the literature review.

11.5.1 Defining Goal and Scope

The application of a combined environmental and social LCA is to compile and evaluate the environmental and social impacts of different life cycle alternatives of a product with a well-defined functionality. For both dimensions the goal and scope definitions should be the same, in order to make impact comparable. Two different inventory analyses are carried out simultaneously followed by a separate impact assessment, typically following the approach of option 1, as stated by Kloepffer (2008).

The functional unit (FU) is the core component of traditional life cycle analysis. It is the common base of fair comparison of different life cycles. Therefore especially in a combined LCA, the FU has a major role, because it has to suit both dimensions. In a traditional LCA by means of the FU, the amount of functionality provided in the life cycle can be scaled on a linear basis. For example, double emissions of

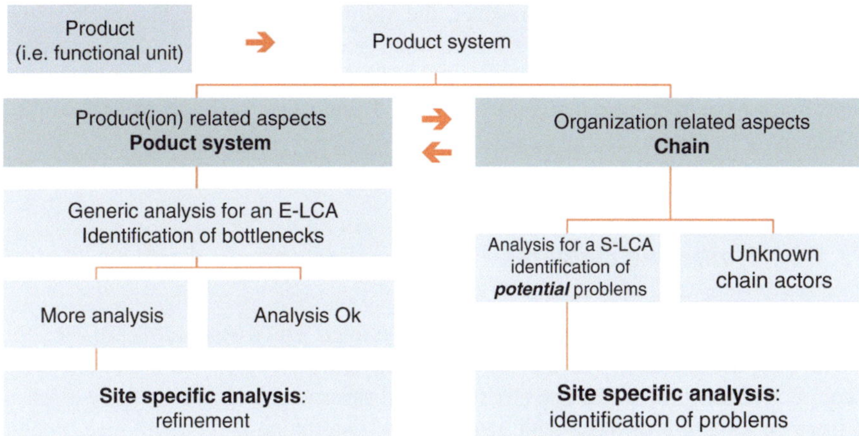

Fig. 11.1 Combined analysis of the product system. (UNEP 2009, p. 38)

CO_2 are likely to result in a doubled impact on the effect of climate change. For social inventory this relation is not always present. For example, 'the grades of customer satisfaction on a scale of 1-10' is not likely to be linearly affected by a change in amount of function required to fulfil the FU. In order to avoid this problem, there are different ways of inventorying social data. One of them is the use of numerical data on social mechanisms such as social metrics as Hunkeler (2006) did. Another is the use of data which is semi-qualitative in nature. The use of numerical data is compatible with the aggregative nature of traditional LCA. However many of the subcategories, as defined in the UNEP guidelines, cannot properly be measured on a quantitative scale. Therefore it is determined to base the SELCA method on the latter method of qualitatively assessing social conduct and by converting this assessment into a quantitative impact profile. The approach for performing this assessment is illustrated in Sect. 11.6.2.

Besides the functional unit, the type of data also influences the level of detail of the LCA. With broad and general data, one can perform an LCA on a broad and indicative level. For environmental LCA, this indicative level means hotspots can already be defined. The level of product detail and the level of inventory together determine the depth of study of an environmental LCA. A broad and indicative level for the social dimension means one can only identify potential problems. In order to specifically identify the problems in the social dimension, one needs detailed site-specific data (see Fig. 11.1). In order to know the type of data that is required, the level of detail of social conduct classification has to be determined upfront. Different levels of detail are (i) sector in a specific region, (ii) industry in a specific region, (iii) organisation-wide or (iv) on site at a specific company or company division, which is, for example, the specific facility responsible for the specific life cycle process.

The validity of the study is closely related to the data requirement and quality. Inventory data in an LCA database can be used by a wide range of practitioners and studies with different scopes. However, the validity of this data, and thereby the validity of this study, strongly varies between geographic, across industries, over time and even between experts. Therefore all limitations encountered and assumptions made have to be systematically recorded. Their effect on the scope and validity of the study has to be carefully considered.

11.5.2 Gathering Data and Performing Inventory Analysis

In the inventory analysis phase, the data are obtained which afterwards are used for the impact assessment. The aim of the SELCA is to be able to simultaneously acquire inventory for social and environmental assessment. However, there exists a difference in nature of these data, respectively, qualitative and quantitative. Social inventory data can be very susceptible to improper or wrong interpretation in spite of accurate descriptions and guidelines. Environmental inventory is mostly based on predefined pathways and has clear indicators which are comparable across process technologies and sectors. By contrast social conduct can vary widely across industries, individual organisations, geographical regions and timeframes. Also, social data can change because of sudden policy changes in social conduct of companies, wherefore system boundaries must be well documented in the goal and scope definition. In social inventory analysis, many implicit assumptions are made, which therefore have to be recorded accurately to reduce the risk of invalidating the impact assessment.

For environmental inventory the interventions are the 'effects' that are caused by 'substances' crossing the system boundary as predefined in impact assessment methods such as ReCiPe[3] (Goedkoop et al. 2009). These environmental interventions can be derived directly from the process tree[4] and are documented in the list of interventions.

The set of social interventions that is included in the SELCA method is defined by social research methods (literature review, interviews to experts and discussions groups). The performance of a social inventory analysis by obtaining social conduct from the associated enterprises, regions or countries is more complex. For every step in the process tree within the cut-off criteria, all relevant social topics/social conduct is indicated. This inventory is documented by the use of the scorecards, where the qualitative social conduct on every social topic is converted into quantitative data. The scorecards define all relevant social topics per process step. These cards have been defined in collaboration with professionals from the textile sector,

[3] ReCiPe is a methodological approach to quantitatively assess the environmental impacts of the life cycle of products, transforming the long list of life cycle inventory results into a limited number of indicator scores.

[4] The process tree refers to the stepwise description of the value chain needed to manufacture any product. Usually it is represented in a flow diagram which can take the tree form.

i.e. experts of the centre 'Texperium',[5] who were consulted to constitute the score-cards for the different social subcategories. For each social topic, a predetermined description of the social mechanism (UNEP 2009) is followed by the selection criteria for its conduct evaluation.

The conduct evaluation for every social topic is classified into one of the five performance indicator classes. These classes are determined on the basis of performance reference points and are defined to be *ideal*, *positive*, *in accordance with international standards*, *negative* and *unacceptable*. For some social topics, these reference points (selection criteria) were already specified in the *Handbook for Product Social Impact Assessment* by PRé Sustainability (Fontes 2016). The others have been formulated by researchers at the University of Twente.

The level of detail of the inventory analysis must match the determined level of detail of the LCA study. A specific-site evaluation provides an indication on what is happening at the specific enterprise and what requires a single performance classification, whereas an entire industrial sector cannot be classified under one performance but requires a broad conduct profile. This is done by indicating the *risk* of occurrence of the different conducts in an entire sector, industry or geographical region.

11.5.3 SELCA Impact Assessment Method

When an enterprise is evaluated on its environmental impact and its conduct towards the relevant social topics, these values have to be characterised. This step is part of the impact assessment phase, where the environmental and social dimensions are separated. For the environmental dimension, the ReCiPe impact assessment method is applied (Goedkoop et al. 2009). For the social dimension, the SELCA SLCIA[6] method was developed.

Per social topic (social intervention), a set of Conduct Characterisation Factors (CCF) is used to determine the quantitative differences between the conduct performance categories. This set can differ per social intervention according to the sector. Sets of CCFs can have a linear as well as a nonlinear scale, to distinguish between relative impact contributions. Table 11.1 is here given to illustrate the CCFs. Even though these sets are inherently arbitrary, they can be used to distinguish between, or emphasise positive and negative impacts. Note that a value of zero for the category 'in accordance with international standards' inherently is equal to a non-present intervention.

The different sets of weighing factors can be determined using the same approach as in LCA (multi-criteria, expert opinion, panel discussion, ranking, etc.). All 31 social impact categories have a Social Topic Weighting Factor (STWF) assigned. Per stakeholder group the weighted social topics are added up to an aggregated midpoint score. Similarly, the scores of the five stakeholder groups can be aggre-

[5] Textile (Open) Innovation Centre based in the Netherlands

[6] SLCIA stands for social life cycle impact assessment.

Table 11.1 Two examples of Conduct Characterisation Factors sets for SELCA (own contribution)

Conduct performance	C.C.F. set A (linear)	C.C.F. set B (nonlinear)
Ideal	+2	+100
Positive	+1	+10
In accordance with international standards	0	0
Negative	−1	−10
Unacceptable	−2	−100

gated to a single endpoint score. If a certain stakeholder group needs more emphasis, a set of Stakeholder Group Weighting Factors (S.G.W.F.) can be applied. It is also possible to aggregate conduct interventions according to their performance indication. The conduct performance profile can be used to identify the overall performance of the entire life cycle, regardless of the stakeholder group or social topic.

In summary, the SELCA impact calculation goes through the four stages (see Fig. 11.2). The first stage is the qualitative classification of company conduct for each social topic, weighted with CCF. The second stage consists of the aggregation of all conduct for each social topic, weighted with STWF, if desired. The third stage aggregates all social topics related to the different stakeholder groups, again weighted using STWF. The final stage consists of the aggregation of all stakeholder groups into a single score for social impact. Alternatively, all conduct interventions can be aggregated directly for profiling social impact regardless of social topic or stakeholder group.

The impact pathway for the SELCA S-IAM[7] is designed in such a way that the most 'subjective' step of conduct classification is concentrated at the start of the impact pathway. This means that the conversion from a qualitative judgement into a quantitative number is performed as early as the inventory stage instead of later stages like in environmental LCA or other methods for social LCA. The anticipated advantage of this choice is that considerations about objectivity can be addressed at the inventory phase instead of the profiling phase.

11.6 The Jeans Case Application on SELCA Method

This study serves as a showcase on the implementation of the SELCA method and to evaluate the proposed methodology on its application to a case with a global value chain. The analysis is limited to four product life cycle scenarios and aims to evaluate the potential, social and environmental problems in the jeans case. For the sake of time and for this study being exploratory, it was crucial to limit the scope of the study. Focus has been put on the regions with stakeholders that have the largest contribution to the global jeans value chain. The results of the combined analysis of the four different recycling scenarios of jeans are discussed in the last section.

[7] S-IAM means here social impact assessment method.

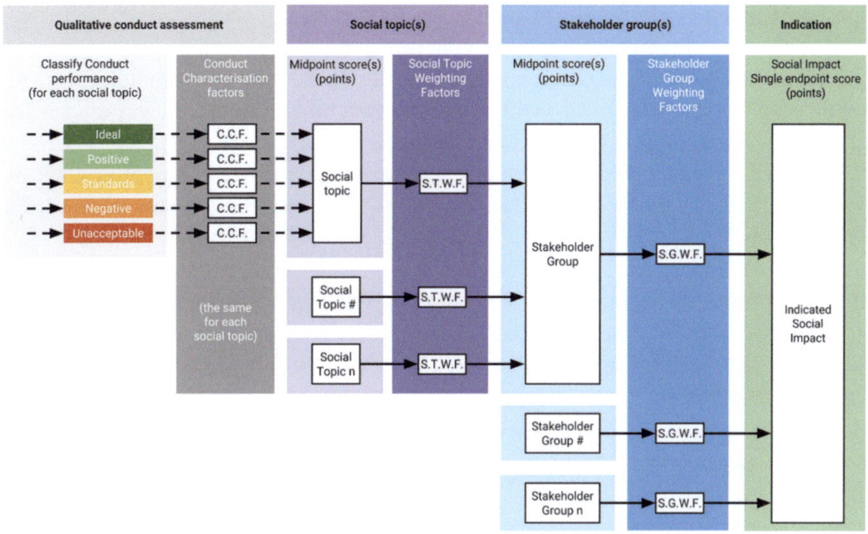

Fig. 11.2 SELCA S-IAM impact calculation pathway (own contribution)

The jeans case has been chosen for two reasons: firstly, because of its global value chain which consists of a wide variety of resources and actors over the product life cycle and, secondly, because of the urge to contribute to the battery of assessment tools to identify the social and environmental impacts of the global textile value chain. Especially with regard to the social dimension, many pressing issues exist. Product lines are refreshed more frequently, life time of clothes reduces and consumption increases. Altogether, this leads to supplier in developing countries to reduce cost and lead times, what consequently worsens the labour conditions in those countries.

11.6.1 Definition of Goal and Scope of Jeans Case

The jeans case explores four different recycling scenarios. Scenarios with partial circularity[8] of materials are compared to a scenario that only uses virgin materials. Here, circularity of materials, what is environmentally in nature, also embraces a social dimension; wherefore, the life cycle assessment of its sustainability can be more holistic.

The functional unit of this case study is defined as: 'To wear denim jeans, bought and used in the Netherlands, 5 days a week, for one year'. All jeans are assumed to

[8]Application of either recovered or recycled materials in the manufacturing phase of new products

have a life span of 10 months of typical wearing, meaning washing every 2.5 days. The depth of study specifies the following life cycle stages: cotton cultivation, yarn production, textile production, jeans production, consumer use and end of life. As cotton can be supplied from many different parts of the world, based on market share and data availability, it is assumed that the cotton cultivation is done in China, where after it is shipped to Bangladesh and America for the subsequent production processes. The consumer use phase and end-of-life phases are restricted to the Netherlands.

As priorly mentioned, four different manufacturing scenarios are used to explore different processes for the production of jeans. The final assessment compared four scenarios (based on materials selection and process conditions) to produce denim fabric: (i) jeans made with 100% virgin cotton, (ii) jeans made with 40% recycled cotton (mechanically recovered), (iii) jeans with recycled PET and (iv) jeans produced with recycled nylon 6, which requires chemical recycling. Regardless of the scenario, a mass ratio of 2:1 of white to blue yarns is assumed for the denim fabric.

11.6.2 Inventory Analysis

The inventory analysis for the environmental part of this combined LCA does not differ from an inventory analysis for a single LCA and therefore is not further elaborated on. As mentioned in Sect. 11.5.2, the inventory analysis for the social LCA is a phase that has to be carried out carefully. Environmental data was sourced from the educational database of the *GaBi* software by the company *Thinkstep*, supplemented with data from the *EcoInvent* database, and social data was based on reports on the social responsibility of the textile sector. Even further, a lack of validated sources for social data has been observed, which can be associated to transparenty issues through the textile value chain. Therefore additional expert knowledge from the centre 'Texperium' has been used. These experts have scored the social issues from different industries and regions on one of the five impact classes.

The end of life of a product is essential for a circular flow of materials and therefore of high importance with respect to the four scenarios. Four end-of-life destinations of the product have been modelled: cloth reuse (10%), garment reuse (7%), new yarn production (14%) and incineration of textile (69%). Finally, the life cycles of each of the four scenarios were simulated. In order to illustrate the needed processes in the cotton-to-jeans life cycle in Fig. 11.3, the scenario of jeans with 40% recycled cotton fibres is presented.

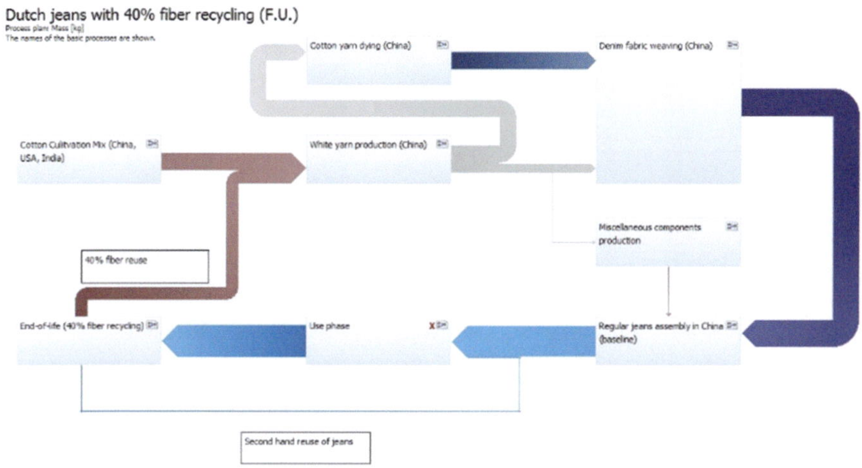

Fig. 11.3 Cotton-to-jeans lifecycle, scenario with 40% recycled cotton fibre

Table 11.2 Example on intervention and characterisation calculation for sector X and company Y

	Intervention profile			Characterised profile	
Performance	Sector X	Company Y	C.C.F.	Sector X	Company Y
Ideal	0.1	0	+2	0.2	0
Positive	0.3	1	+1	0.3	1
Standard	0.4	0	0	0	0
Negative	0.2	0	−1	−0.2	0
Unacceptable	0	0	−2	0	0
				0.3 pts	1 pts

11.6.3 *Impact Assessment*

For the environmental dimension, impact is assessed using the ReCiPe v1.08 method (Goedkoop et al. 2009). In order to illustrate the method for the social dimension, an example is given in Table 11.2 for the intervention 'fair salary conduct' of sector X and company Y. Two different entities have been used to show the possible difference in level of detail. Note: this example is separate from the 'jeans' case. A score profile for both entities has been determined during inventory analysis. Next, this profile is characterised via its set of CCFs, which can differ per case. This results in an indicator score for intervention.

Together with the other intervention scores, this score is aggregated into *midpoint* scores (see Fig. 11.2) for the social topic or for the five stakeholder groups. The overall performance endpoint can be calculated by counting the relative presence of conduct performance between the social mechanisms.

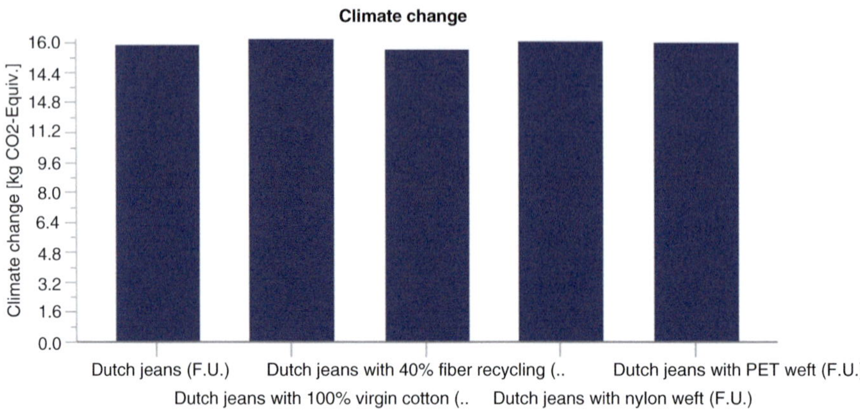

Fig. 11.4 Impact on climate change effect of the baseline and four scenarios

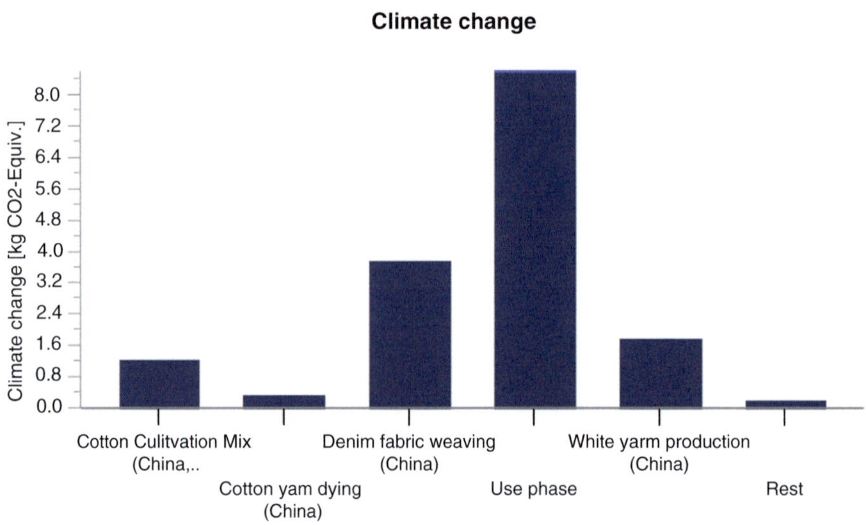

Fig. 11.5 Impact on climate change effect of life cycle phases

11.6.4 Profiling and Evaluation

The environmental impact profiles of the different scenarios show little differences. To illustrate this, the profile of the climate change effect is displayed in Fig. 11.4. Whilst in Fig. 11.5, the impact profile of the average Dutch jeans over the life cycle is indicated. It can be observed that the use phase, the washing, drying and lifespan, causes the highest environmental impact through the whole life cycle of jeans.

The social impact assessment reveals a profile, displayed in Fig. 11.6, where poor worker health and safety have the worst impact, especially during jeans assem-

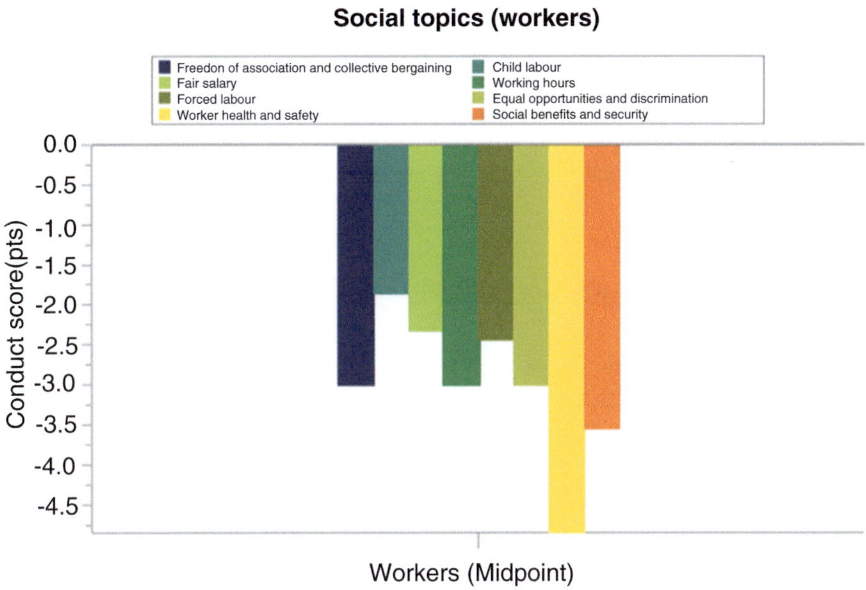

Fig. 11.6 Conduct score of average Dutch jeans on social topic 'workers'

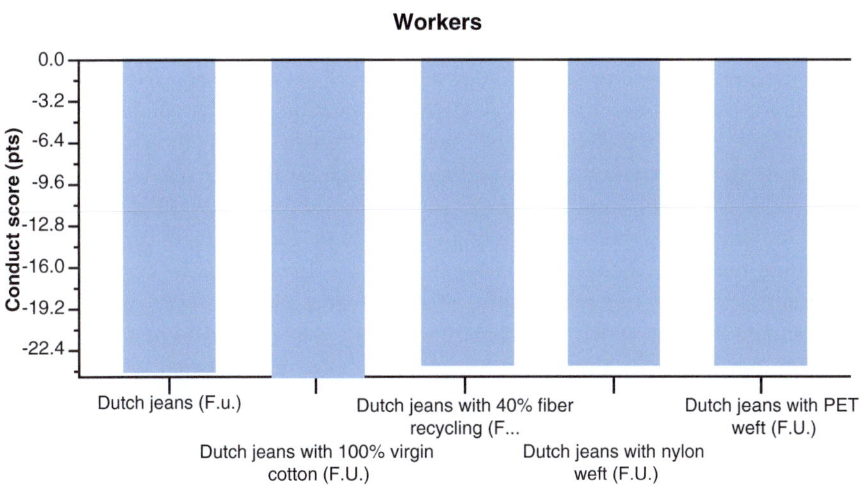

Fig. 11.7 Impact of different scenarios on social topic 'workers'

bly and cotton cultivation. Even further, the impact of the different scenarios on "workers" does not show any significant difference among them (see Fig. 11.7), only a slight bigger impact on the scenario of the Dutch jeans with 100% virgin cotton as functional unit (FU).

When looking at the three scenarios of fibre recycling, for climate change impact, the use of recycled cotton fibres can be considered as the best option, as well as for water depletion. The scenario of recycled nylon fibres proves to be the worst on both impact categories, for climate change, it is even worse than the regular jeans scenario. The social profile shows that the jeans using recycled cotton fibre scores are slightly less bad with −23.4 points compared to −24.4 points for the jeans with only virgin cotton, according to calculations of *GaBi* software used. However, this is a result of scaling and allocation rules by use of the functional unit.

11.7 Conclusions and Recommendations

This section summarises and concludes both the subject matter of the combined social and environmental LCA (SELCA) method and the practical application of this method in a case study on jeans. The research question driving this research (*How can social aspects be integrated into (traditional) environmental life cycle assessment by using denim jeans as showcase?*) was formulated trying to respond to the 'how' question. Considering this type of research, the following aspects were the most prominent.

Due to the infancy of the field, existing approaches for combined social and environmental LCA were not readily available. As existing 'traditional' LCA software is based on linear input-output calculations, they do not allow for 'qualitative' interventions. Based on the current state of the art in LCA literature and best practices (such as the UNEP recommendations), the SELCA approach was developed. SELCA allows for a semi-qualitative assessment of social impact, which makes it suitable for use in conventional LCA software and allows for the combined assessment of both social and environmental impact within the scope of a single study.

Despite the novelty and short development cycle of the SELCA method, it did prove to be a valuable tool during the case study on the life cycle assessment of jeans for predicting performance and identifying improvement opportunities. Regarding the four evaluated scenarios of the case study, it can be concluded that the material recycling of cotton fibres is a substantial measure with respect to the environmental effects on climate change and water depletion. Additionally, the use of water and electricity during the use phase also has a significant impact on the environment, which is mostly determined by consumer behaviour related to washing and drying of their garment. From a social perspective, the bulk of the social impact originates from the global textile supply chain, where the impact on workers was the largest, especially during cotton cultivation, fabric weaving and garment assembly.

The application of SELCA did present a number of noteworthy empirical challenges. From a practical perspective, the inventory analysis of social aspects proved to be difficult due to the lack of transparency in the value chain, which presents a peculiar 'chicken and egg' problem to social LCA. Social inventory also proved to be prone to subjective judgement, despite the careful use of UNEP recommended classification criteria and the use of external documentation during conduct assessment. Time and manpower constraints forced reducing the scope of the LCA analysis to global risk indication due to the labour intensity of gathering social inventory data. During the evaluation of the case study, it was found that the classification of 'standard', 'negative' and 'unacceptable' were prevalent in the impact profiles. Additional case studies could determine if the effect is inherent to this method or if it is symptom of this case study.

A different set of challenges were encountered from a theoretical standpoint, most noticeably, the use of the functional unit (FU) and its role in allocating social impact. Where the FU is designed to linearly scale and allocate environmental impact based on the provided functionality, the same does not necessarily apply to (semi-)qualitative matters such as social impact. This issue is most noticeable in the inclusion of circular principles. In the case study, both recycling and reuse lead to a reduced environmental footprint per FU, but the question remains if these measures should also reduce social impact accordingly. Another challenge was experienced in utilising a consistent scope, system boundaries and cut-off criteria. For some processes in the product life cycle, there was an overlap in environmental and social concerns, as observed in the study in the global textile supply chain. In other occasions, the system boundaries were very dissimilar between environmental and social aspects, as observed in the assessment of the use phase.

The case study on the inclusion of the social dimension into LCA has demonstrated that it is possible to generate a number of valuable insights and preliminary conclusions. However, there remains a need to verify and elaborate on these findings in future research. The foreseeable next step is to evaluate this new approach using an expert panel of various stakeholders in the textile sector.

This study aims to be the starting point of the development of a methodology that contributes towards an integrated assessment on the social, environmental and economical dimension of the life cycle of a product system.

Acknowledgements Special acknowledgement goes to Mr. Peter Bos and Ms. Paula Konter (Texperium Textile Innovation Centre in Haaksbergen, The Netherlands) who generously provided information about the textile industry. They commented on the midterm and final project reports, enriched discussion meetings and enabled the execution of this project. Besides, acknowledgement goes to Ms. Paula Steenstra for combining several sources and previous internal documents into this chapter. Her literature study contributed to this chapter.

References

Arvidsson R, Baumann H, Hildenbrand J (2015) On the scientific justification of the use of working hours, child labour and property rights in social life cycle assessment: three topical reviews. Int J Life Cycle Assess 20(2):161–173. https://doi.org/10.1007/s11367-014-0821-3

Barbier EB (2012) The green economy post Rio+20. Science 338(6109):887–888

Baumann H, Arvidsson R, Tong H, Wang Y (2013) Does the production of an airbag injure more people than the airbag saves in traffic? J Ind Ecol 17(4):517–527

Benoît-Norris C, Vickery-Niederman G, Valdivia S, Franze J, Traverso M, Ciroth A, Mazijn B (2011) Introducing the UNEP/SETAC methodological sheets for subcategories of social LCA. Int J Life Cycle Assess 16(7):682–690. https://doi.org/10.1007/s11367-011-0301-y

Benoit-Norris C, Cavan DA, Norris G (2012) Identifying social impacts in product supply chains:overview and application of the social hotspot database. Sustainability 4(9):1946
Chhipi-Shrestha GK, Hewage K, Sadiq R (2015) 'Socializing' sustainability: a critical review on current development status of social life cycle impact assessment method. Clean Techn Environ Policy 17(3):579–596. https://doi.org/10.1007/s10098-014-0841-5
Ciroth A, Franze J (2011) A comparison of cut roses from Ecuador and the Netherlands. Int J Life Cycle Assess 16(4):366–379
Dreyer L, Hauschild M, Schierbeck J (2006) A framework for social life cycle impact assessment (10 pp). Int J Life Cycle Assess 11(2):88–97. https://doi.org/10.1065/lca2005.08.223
Dreyer LC, Hauschild MZ, Schierbeck J (2010) Characterisation of social impacts in LCA. Int J Life Cycle Assess 15(3):247–259. https://doi.org/10.1007/s11367-009-0148-7
Elkington J (1994) Enter the triple bottom line. In: Cannibals with forks: the triple bottom line of 21st century business. Capstone Publishing Ltd, Oxford
Finkbeiner M, Schau EM, Lehmann A, Traverso M (2010) Towards life cycle sustainability assessment. Sustainability 2:3309–3322. https://doi.org/10.3390/su2103309
Fontes J (2016) Handbook for product social impact assessment: roundtable for product social metrics version 3.0. PRé sustainability
Foolmaun RK, Ramjeeawon T (2013) Comparative life cycle assessment and social life cycle assessment of used polyethylene terephthalate (PET) bottles in Mauritius. Int J Life Cycle Assess 18(1):155–171. https://doi.org/10.1007/s11367-012-0447-2
Goedkoop M, Heijungs R, Huijbregts M, De Schryver A, Struijs J, Van Zelm R (2009) ReCiPe 2008. A life cycle impact assessment method which comprises harmonised category indicators at the midpoint and the endpoint level 1
Guinée J (2016) Life cycle sustainability assessment: what is it and what are its challenges? In: Clift R, Druckman A (eds) Taking stock of industrial ecology. Springer International Publishing, Cham, pp 45–68. https://doi.org/10.1007/978-3-319-20571-7_3
Hunkeler D (2006) Societal LCA methodology and case study (12 pp). Int J Life Cycle Assess 11(6):371–382. https://doi.org/10.1065/lca2006.08.261
ISO (2006) ISO 14040. Environmental management – life cycle assessment – principles and framework. ISO, Genova. doi:citeulike-article-id:4482380
Jørgensen A, Le Bocq A, Nazarkina L, Hauschild M (2008) Methodologies for social life cycle assessment. Int J Life Cycle Assess 13(2):96. https://doi.org/10.1065/lca2007.11.367
Jørgensen A, Herrmann IT, Bjorn A (2013) Analysis of the link between a definition of sustainability and the life cycle methodologies. Int J Life Cycle Assess 18(8):1440–1449
Kloepffer W (2008) Life cycle sustainability assessment of products. Int J Life Cycle Assess 13(2):89. https://doi.org/10.1065/lca2008.02.376
Los B, Timmer MP, de Vries GJ (2015) How global are global value chains? A new approach to measure international fragmentation. J Reg Sci 55(1):66–92. https://doi.org/10.1111/jors.12121
Macombe C, Leskinen P, Feschet P, Antikainen R (2013) Social life cycle assessment of biodiesel production at three levels: a literature review and development needs. J Clean Prod 52(1):205–216
Mair S, Druckman A, Jackson T (2016) Global inequities and emissions in western European textiles and clothing consumption. J Clean Prod 132:57–69. https://doi.org/10.1016/j.jclepro.2015.08.082
Mattioda RA, Mazzi A, Canciglieri O, Scipioni A (2015) Determining the principal references of the social life cycle assessment of products. Int J Life Cycle Assess 20(8):1155–1165. https://doi.org/10.1007/s11367-015-0873-z
Norris GA (2006) Social impacts in product life cycles – towards life cycle attribute assessment. Int J Life Cycle Assess 11(1):97–104. https://doi.org/10.1065/lca2006.04.017
O'Brien M (1996) Social and environmental life cycle assessment (SELCA): approach and methodological development. Int J Life Cycle Assess 1(4):231–237
OECD (2004) A new world map in textiles and clothing: adjusting to change. OECD, Paris

Parent J, Cucuzzella C, Revéret J-P (2010) Impact assessment in SLCA: sorting the sLCIA methods according to their outcomes. Int J Life Cycle Assess 15(2):164–171. https://doi.org/10.1007/s11367-009-0146-9

Petti L, Serreli M, Di Cesare S (2016) Systematic literature review in social life cycle assessment. Int J Life Cycle Assess:1–10. https://doi.org/10.1007/s11367-016-1135-4

Ramirez P, Petti L (2011) Social life cycle assessment: methodological and implementation issues. Annals of The "Stefan cel Mare" University of Suceava 11(1):11–17

Schmidt AC, Jensen AA, Clausen AU, Kamstrup O, Postlethwaite D (2004) A comparative life cycle assessment of building insulation products made of stone wool, paper wool and flax. Int J Life Cycle Assess 9(1):53–66. https://doi.org/10.1007/bf02978536

Spillemaeckers S, Vanhoutte G, Taverniers L, Lavrysen L, van Braeckel D, Mazijn B, Rivera J (2004) Integrated product assessment – the development of the label 'sustainable development' for products ecological, social and economical aspects of integrated product policy. Belgian Science Policy, Brussels

Sutherland JW et al (2016) The role of manufacturing in affecting the social dimension of sustainability. CIRP Ann Manuf Technol 65(2):689–712

Taplin IM (2006) Restructuring and reconfiguration: the EU textile and clothing industry adapts to change. Eur Bus Rev 18(3):172–186. https://doi.org/10.1108/09555340610663719

Taplin I (2014) Who is to blame? A re-examination of fast fashion after the 2013 factory disaster in Bangladesh. Crit Perspect Int Bus 10(1/2):72–83

Timmer MP, Los B, Stehrer R, de Vries GJ (2013) Fragmentation, incomes and jobs: an analysis of European competitiveness. Econ Policy 28(76):613–661. https://doi.org/10.1111/1468-0327.12018

Traverso M, Bell L, Saling P, Fontes J (2016) Towards social life cycle assessment: a quantitative product social impact assessment. Int J Life Cycle Assess:1–10. https://doi.org/10.1007/s11367-016-1168-8

UNEP (2009) Guidelines for social life cycle assessment of products. http://www.unep.fr/shared/publications/pdf/dtix1164xpa-guidelines_slca.pdf. Accessed June 2016

UNEP (2011) Towards a green economy: pathways to sustainable development and poverty eradication

Wang SW, Hsu CW, Hu AH (2016a) An analytic framework for social life cycle impact assessment – part 1: methodology. Int J Life Cycle Assess 21(10):1514–1528. https://doi.org/10.1007/s11367-016-1114-9

Wang SW, Hsu CW, Hu AH (2016b) An analytical framework for social life cycle impact assessment – part 2: case study of labor impacts in an IC packaging company. Int J Life Cycle Assess:1–14. https://doi.org/10.1007/s11367-016-1185-7

Weidema B (2005) ISO 14044 also applies to social LCA. Int J Life Cycle Assess 10(6):381. https://doi.org/10.1065/lca2005.11.002

Weidema B (2006) The integration of economic and social aspects in life cycle impact assessment. Int J Life Cycle Assess 11(1):89–96. https://doi.org/10.1065/lca2006.04.016

Wu R, Yang D, Chen J (2014) Social life cycle assessment revisited. Sustainability 6(7):4200

Chapter 12
A Circular Model of Residential Composting in Mexico City

Vivian Plasencia-Vélez, Marco Antonio González-Pérez, and María-Laura Franco-García

Abstract The urban solid waste (USW) in Mexico City is managed at the municipal level. This situation means several challenges: the 3-year municipal administration period clearly affects the continuity of their USW management plans and programmes, adding to the space shortage problem to properly landfill them. Even further, the technologically insufficient operation of landfills represented 16% of greenhouse gas emissions in 2016. Organic waste represented between 45% and 55% of the total USW. Therefore, grassroots initiatives were the focus of this research because some of them proved to reduce USW at the household level, because activities to turn the organic waste into compost by community members are relevant. This fits into the purposes of the circular economy and zero waste landfill. Local composting has an important potential to improve USW management: the goal of this paper was to identify the conditions necessary for those projects to be successful. Hence, our research question is as follows: Which are the conditions needed to facilitate the community-based compost production? To answer it, international cases were analysed to learn from the existing best practices. Two countries were used as reference: the United Kingdom and the Netherlands. A proper literature review was carried out to build the analytical framework to assess one specific case study in Mexico City: the composting plant Club de Golf Bellavista (CGB). Surveys and interviews were carried out in order to compile empirical data and information for further analysis. Among the most relevant findings, "social participation" came across as a relevant factor in this type of grassroots initiatives, particularly at the source generation of the USW. This was consistently mentioned through

V. Plasencia-Vélez (✉)
World Resources Institute, Belisario Domínguez 8, Col. Villa Coyoacán, Delegación Coyoacán, Mexico City, Mexico

M. A. González-Pérez
Faculty for Higher Education (FES), Iztacala Autonomous National University of Mexico (UNAM), Tlalnepantla, Mexico State, Mexico

M.-L. Franco-García
University of Twente, Enschede, The Netherlands
e-mail: m.l.francogarcia@utwente.nl

M.-L. Franco-García et al. (eds.), *Towards Zero Waste*, Greening of Industry Networks Studies 6, https://doi.org/10.1007/978-3-319-92931-6_12

surveys responded by CGB stakeholders. Neighbours also had the opportunity to suggest different mechanisms that could convince other neighbours to engage in the separation phase of USW.

Keywords Compost · Residential areas · Organic waste · Sustainable scheme for local communities

12.1 Introduction

Through the study and benchmarking of two successful USW management programmes in the Netherlands and the United Kingdom and the study case in Mexico City, it was found that by sorting residues, they may be reduced and recycled. This contributes to create a circular economy in the community as well as to achieve the goal of zero waste to landfill.

The waste management in Mexico is a very relevant problem due to the lack of effective mechanisms to reduce, reuse or recycle the discharged materials post consumption and the poor culture to recover materials. This is accompanied by scarce policy instruments that either incentive or force citizens to manage their wastes in a more environmental friendly fashion.

Through the analysis of documents and interviews carried out in Mexico, the United Kingdom and the Netherlands, it is possible to propose a scheme to manage and handle household waste, specifically the organic waste, which represents between 45% and 55% of the USW in Mexico (Álvarez 2002). Organic waste can successfully be recycled by producing compost, which can be easily done by members of local communities.

The compost plant used in this study case, operating since 2008, is located in Bellavista, in the municipality of Atizapán de Zaragoza in Mexico State. Compost produced from the households' organic wastes works as a soil improver, substituting the use of chemical fertilisers in the green areas inside the residential area, which has had reduced the costs for maintenance of green areas. To replicate its process in other residential areas, the scheme needs improvement.

Some similar cases to Bellavista were found in countries like China and Bangladesh, but it was the United Kingdom which had one of the most interesting schemes. There, the community compost plants are a common way for organic waste recycling, and the existence of a community compost network is also analysed as a next step for waste management in Mexico. In fact, social participation was identified as a vital aspect for compost plant operation. If local people do not separate their waste, compost production would be impossible.

The investigation focuses on the collection of organic waste from houses in order to develop a low-cost recycling process by producing compost within a community and create best practices based on the experiences analysed.

12.2 The Problem of Waste Management in Mexico City

Waste management is the responsibility of local authorities or municipalities. In spite of elaborated regulatory environmental frameworks, these are insufficiently enforced due to limited capacities and poorly observed social participation that is reflected in the lack of waste separation. This latter is one of the most important barriers towards an efficient waste management.

The municipal waste is divided into non-perishable or non-biodegradable residues (glass, laminates, plastics, etc.) and organic waste. Waste is regarded as organic if it is biodegradable and can be absorbed, almost naturally, back into the environment. In the case of Mexico, non-biodegradable waste generates "scavenging": a process of selection and classification that allows to take advantage of 5–10% of waste (Herrera 2004) in various industrial processes.

Organic wastes are being underutilised and scattered over large areas of garbage disposal (open dumps), which are permanent sources of pollution. Likewise, there is no recycling culture in Mexico, and there is a shortage of collection centres in order to reuse and recycle waste, either organic or inorganic (Castillo 2002).

It is also known that the collection of waste is not efficient, because of the scavengers' interests. Therefore, there is an irregular collection of waste, and collectors fight over waste zones in the city, preferring areas which pay more to get their waste collected (Bernache 2006).

Needless to say, Mexico City's landfills have exceeded their capacity. Although efforts to recycle have been made to produce compost and generate biogas to create electricity for the municipality of Atizapán de Zaragoza (Medellin 2015), this is not enough when considering the lack of law enforcement and of mechanisms that could encourage citizens to separate their waste.

When residues are only disposed into landfills, the pollution of water reservations also represents a problem. Another consequence is the shortage of space for waste disposal in landfills. On the other hand, this can also be considered as an opportunity for the implementation of community composting programmes in order to reduce organic waste dumped into landfills.

Dumping wastes in landfills is both a public health and environmental risk. Since it is a cycle of environmental pollution, the leachate drains from the slopes or filtrates into the subsoil, and this often has the characteristics of hazardous waste (Restrepo and Phillips 1991). Waste also attracts vermin such as rats, flies, other insects and worms that live and reproduce in large quantities. They foster diseases and, as the wind distributes large amounts of waste in vast areas, damage the natural environment. These reduce the quality of life thus having negative effects in aspects like hygiene, health and public welfare (Dennis et al. 2006). Society puts pressure on the natural system, and finding options to restore an appropriate interaction between the actions of the social and natural system becomes necessary (Vicente and Reis 2008).

Additionally, composting in local conditions impacts positively on the operation of the municipality and is valued internationally because of the greenhouse gas reduction. This has been supported by the Mexico City government through its Environmental Ministry (SEDEMA, for its Spanish acronym) since 2006. Greenhouse gas emissions in residential areas in Mexico City constituted 9.8% of the total emissions. The biggest landfill in Mexico City is *Bordo Poniente*, which receives approximately from 4,380,000 to 5,110,000 tons of waste and emits 2 million tons of carbon dioxide into the atmosphere per year (La República 2008). In Mexico City, USW represents 16% of greenhouse emissions (Mejia 2016).

In this research, the landfill *Puerto de Chivos* was studied due to its proximity to the area of the case study: the Compost Plant in the Residential Area in *Bellavista*. In that landfill, 500 tons of wastes are disposed, although it has been restricted since 2007 because of its limited space (Medellin 2015).

The proposed model for waste management is the implementation of compost plants in residential areas, but in order to do this, it is necessary to analyse the drivers and barriers of the implementation of a community compost plant. These include legal, economic, operative, environmental and social variables. Through the analysis of different case studies from Mexico (MEX), the Netherlands (NL) and the United Kingdom (UK), we suggest best practices and ways to improve the current residential scale model for the *Club de Golf Bellavista* compost plant.

12.3 International Benchmarking Experiences

The two international studied cases (United Kingdom and The Netherlands) are here further described to identify some of the best practices of social participation in waste separation and compost production.

12.3.1 British Case

The United Kingdom is a good example when it comes to observing the best mechanisms to manage waste. Even though they have successful industrial processes for managing waste, there are also local communities' initiatives that cannot be overlooked. These latter are what the research focuses on.

In the United Kingdom, community compost production has developed quickly, and the successful cases are many. Their main objective was to reduce waste from landfills or incineration, thereby preserving internationally important peat land natural habitats and improving the topsoil structure.

Five compost plant's managers answered an exploratory questionnaire with the objective of knowing the incentives and barriers for the implementation of this scheme. These compost plants were the Scottish Composting Centre, Brighton Community Compost Centre, Lower Slaughter Community Composting, Cwm Harry Land Trust and Denton Parish Council.

One of the main motivations to create and implement community compost plants is the international pressure, which leads to the implementation of legal incentives. Another incentive is the money savings that come from the produced compost, because it can be used as fertiliser for the green areas instead of buying chemical fertilisers.

The independent variables contemplated for this study are policy incentives, project management (technology used, financial situation and operation of the plant), manager's perception (which can be translated into the creation of networks) and society's environmental conscience (which can be translated into participation). This helped to know more about the operation of the project and its organisation, the technology used, the financial situation and stakeholders, the environmental impact, management and success factors. The management is bottom-top, because it started from a social initiative and has been growing through the creation and strengthening of community compost networks.

The level of success regarding the implementation of compost plants in the United Kingdom is influenced by the variables described in Fig. 12.1, which were mentioned by the compost plant managers when they were interviewed.

Success includes aspects such as adequate infrastructure, quantity and quality of compost, investment and feasibility of the programme. These aspects were relevant at the first operational stage of the implementation of composting plants, and they are analysed as possible barriers.

When the waste disposed into the landfills is reduced, it has a positive effect in the environment because less waste is transported to final disposal locations and compost is produced near the end user. In the long term, waste could work as a bio-input to generate electricity, and the residues could be used to produce compost. The economic feasibility of the project does not represent an obstacle, because some of the compost plants, like the Denton Community Compost, have the local government support.

One of the most common barriers perceived by the community compost plant managers is that compost production takes a long time (about 80 days), and, in order

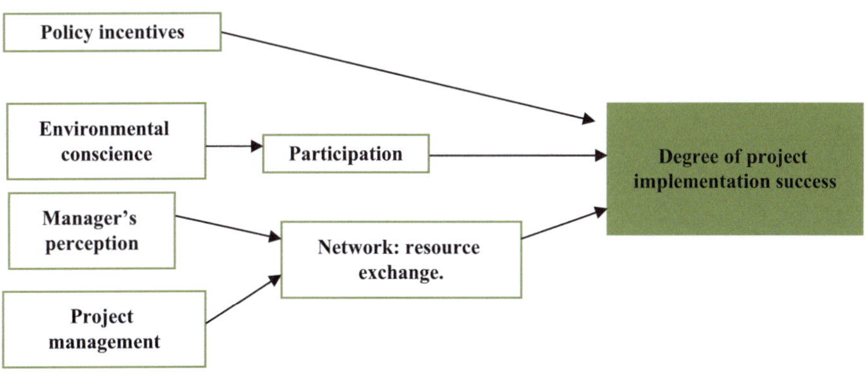

Fig. 12.1 Degree of implementation success of community compost in the United Kingdom. (Own contribution)

to sell the compost, it has to be certified by the "British Standards Institution" (2005) "Publicly Available Specification for Composted Products" (PAS 100), certification by the Composting Association and the "EU Eco-label for Soil Improvers and Growing Media" certification by the Department for Food and Rural Affairs. Among other certifications, there are the "Organic Standards for Landscape and Amenity Horticulture" certification by Henry Doubleday Research Association (HDRA) and the "Soil Association Standards for Organic Food and Farming" certification by the Soil Association (Slater 2004).

These certifications are not mandatory, but in Scotland, for example, in order to be able to claim that waste has been fully recovered, the process/product must meet a nationally recognised standard or specification. If a product does not meet such a standard or specification, the local authority is not able to count this waste towards their statutory targets.

Community composts are allowed in the Environmental Permitting Guidance, Waste Framework Directive which states that households do not require permits for carrying out waste operations involving only household waste or managing of waste within their own property. Disposal of waste by householders is not allowed when it is likely that it causes harm to the environment or human health (Environmental Agency 2008).

The community compost plants analysed in the United Kingdom process organic waste from households, particularly from small communities with maximum 50 households. Waste is processed manually and, in average, it takes 2 months to obtain fertiliser.

If fertiliser cannot be sold, that endangers the feasibility of the programme. Therefore, communities need an environmental permit to start the compost plant and hire people with technical expertise in the production of compost, because if it is not handled in a professional way, it could be potentially dangerous to the environmental and human health.

The United Kingdom scheme is considered a positive practice in this research because it is a bottom-up policy, which could be more likely to be implemented in the Mexican case. The lack of implementation of the top-bottom waste management policy because of stakeholders' interests (scavengers) could create the opportunity for the implementation of private programmes where waste is recycled through community compost, creating electricity or heat.

12.3.2 Composting Network in the United Kingdom

There are several associations and organisations that encourage the creation of community composts through a bottom-up policy which involves direct participation of citizens in the recycling of their organic waste. There is even a community compost network which provides advice and training to its members (Organic recycling 2010).

An example of this is the Association for Organics recycling, (Organic recycling 2010) formerly Composting Association, created in 1995, which promotes the sustainable management of biodegradable resources. The Association encourages a regulatory and economic framework suitable for the long-term benefit and sustainability of the composting industry as a whole. It also centralises information and researches and distributes them.

Lately, the Composting Association has over 700 members from all sectors of the UK waste management industry, including compost producers, regulators, local authorities, consultants, trade suppliers, compost users, academics, individuals and students.

Its purposes are to promote the environmental and economic benefits of effective composting and compost use, promote a quality standard for composts (which includes requirements covering compost production and monitoring), sampling and test methods, compost quality and declaration/application information. The aims of standards (BSI PAS 100:2005) are to assist producers in maintaining consistent and reliable products and to encourage greater consumer confidence in composts. The Composting Association manages the certification scheme (Association for Organic Recycling 2010).

Another example is the Community Compost Network, which has over 230 members who receive a quarterly newsletter and have access to The Growing Heap website, e-mail news list and directory of members, annual conference and training events.

This organisation shares some of the objectives with the Composting Association, such as the development of biological treatment as a sustainable waste management technique, and develops and runs workshops, seminars and training events to improve the skills and knowledge of people working within the composting industry, regulatory agencies, local authorities, professionals in related industries and the general public. It also works with consultancy services (Community Compost Network 2010). In order to be part of any of these organisations, there is an annual fee depending on the characteristics of the applicant.

This particular scheme works because of the lack of industrial production of compost from organic waste. This could be used as a good example for the Mexican case.

12.3.3 Dutch Case

Waste management in the Netherlands has proven to be within the highest European standards, and for such reason, it is considered one of the role models at this regard. This started since 1960, when environmental pollution represented an issue to this country. When the government wanted to act on it, legal instruments did not exist. The government opted for a sector legislation being enacted for each type of pollution. This was not effective because the legislation was too detailed and coordination was inadequate.

In 1990, it was decided that this Act should be transformed into a single integrated piece of environmental legislation. This resulted in the Environmental

Management Act, which came into force in March 1993. This Act stipulates rules regulating the discarding and collection of waste and on its processing, reprocessing, destruction and final disposal.

One of the biggest contributions of this Act is the establishment of rigorous standards for landfill sites and incinerators. Also, the landfill tax tends to make these ways of disposal expensive, thereby encouraging different ones, such as compost plants. Furthermore, the landfill of many types of waste is prohibited (VROM 2010).

Legislation becomes even stricter as years go by, due to the compromises acquired through international protocols. This is why the Third National Environmental Policy Plan set an indirect target for prevention: the growth of waste generation over the period 2000–2010 should be 20% less than the growth of the economy (VROM 2010). The Environmental Management Act regulates the management of hazardous residues, domestic waste, emissions to the air, landfill management and their possible prohibitions, batteries disposal, the car tire decree and the waste oil decree.

There is also a chapter which specifically addresses waste and stipulates the hierarchy of its management: prevention, reduction, reuse, recycle, incineration and finally landfill. Ever since January 1, 1994, the Act mandates that local authorities collect household waste, separating it into organic and inorganic. The organic waste includes garden and kitchen waste, which is processed to produce compost and then used as soil improvers. The provincial environmental management requires local authorities to organise the collection of paper, cardboard, glass, textiles and chemicals away from their source of generation.

The citizens' organic waste is collected once a week, the next week the inorganic waste is collected and so on. Another successful aspect of the Dutch legislation is the green tax, implemented in 1994, for using the landfill. The objectives of this tax are to provide a revenue source to the local government and to have a positive impact in the environment. The fact is that the costs of landfills are lower than using the incinerator, which is why the government decided to increase the price to deposit waste in the landfill. Since the cost to incinerate is lower, people are incentivised to separate their garbage to be used for incineration and compost production.

In many cities, this tax is calculated according to the weight of the garbage or the number of times a garbage can is presented for emptying. The revenue raised by this tax was estimated at 117 million Euros in 2005 (VROM 2010).

Taking this into account, a case from the region of Twente, which is located in the province of Overijssel, eastern part of the Netherlands, was studied. This region has three large cities: Hengelo, Enschede and Almelo, which have 77,500, 150,000 and 65,200 inhabitants each one (Business and Science Park 2010). The focus of the research was mainly in the city of Hengelo, where an exploratory questionnaire was applied in the format of an interview to the person in charge of the composting process in Twence, which is a semi-governmental waste treatment company. The interview was divided in six categories: operation of the organisation, technological innovation and product characteristics, financial situation, stakeholders, environmental impact of the project and management (Medellin 2015).

The region of Twente has a company which is responsible for the waste management called Twence. Since 1986 it has published a sustainable vision, when a landfill and a station for the disposal of hazardous household waste were opened. At that same site, a composting plant began operating in 1994. It was in 1997 when they began selling their product.

At Twence, the entire process of composting organic waste from kitchens and gardens takes place in sealed spaces, using the so-called closed method, which is an aerobic method. The compost is an excellent soil enhancer for agriculture and horticulture, and it is certified by the Authority of Food Safety and Consumers (VWA-erkenning, for its Dutch acronym), which verifies that the compost plant operates under legal requirements of the norm 1774/2002, which allows the mixing of animal byproducts with organic waste that has been used for compost, whose source is gardens and home kitchens, used under hygienic conditions.

The certificate, BRL [National Beoordelingsrichtlijn] GFT Compost, is specifically for the compost produced from organic waste that comes from kitchens and gardens. The certificate BGK [Bundesgütegemeinschaft Kompost e.V.] allows the sale of compost produced in Twence to Germany.

There are two end products: green compost (consisting of elements of 0–10 mm in size) and coarser wood components (larger than 10 mm in size). The latter material can be used as a secondary fuel, intended to be used as such in the biomass power plant. By means of this process, Twence is able to treat 30,000 tons of green waste each year (VROM 2010), and over 90% of the waste received at Twence is transformed into raw materials, building materials and energy (Twence 2010).

The process is feasible because Twence receives money for processing waste. In the case of the incineration of waste, the electricity and the heat produced are sold to inhabitants of the region or other countries. In the case of compost, it is given away for free to nearby farms. This scheme is environmentally friendly, and Twence takes into account the neighbours and the inhabitants by avoiding odours in the waste treatment. That also works as a strategy to motivate people to separate their waste (Twence 2010).

As far as social interaction, there are also reported cases of citizens' complaints about unpleasant odours and noise, litter, excessive traffic and nuisance from birds; these are indicators that reflect aspects that could upset inhabitants, and the plant takes them into account each year in order to improve its performance.

In the framework of the plan set out in 2006 to further reduce the odour impact of the plants, they have worked on better timing of the turning of the composting materials and overflow with regard to the weather conditions, such as the direction of the wind. At the post-composting phase, a door has been replaced with a so-called quick-lock door, whereby less emission of the process air is achieved. Finally, the compost produced from organic waste separation is used by farmers as soil improvers.

By analysing the situation of waste management as a project in a local scale in the Twente region, it is easier to understand what are the drivers and barriers for the implementation of this policy, which could represent a successful model. The success factors are understood when variables such as the project's environmental impact, stakeholders' participation, financial aspects, legal aspects, technology used and the operation of the compost plant are taken into account.

These variables were considered in a questionnaire, which was applied in an interview with the Twence compost plant manager (Tcpm). He implied that the governments' decision to change the public policy on waste management began with the lack of space to landfill the waste. When this issue became more relevant, the Netherlands' government approved a strict regulation on waste management, where every household has to separate their waste into organic and inorganic (Ministery of Housing, Spatial planning and the Environment 2001) in order to process it more efficiently into electrical energy, heat and/or compost. That is why this scheme considers managing the organic waste in an industrial level, instead of a small-scale community compost plant.

In the region of Twente, there is no competition in the processing of waste and production of compost. In the words of the Tcpm:

> We live in the east of the Netherlands and here is no problem as people separate the trash. We have the potential to grow because our raw material is constant and now we process waste from Germany and other parts of Netherlands. Since 2001 we opened the plant to treat waste from Rotterdam and Amsterdam and since 2008 from Germany too.

The first step in the process to produce compost starts when the organic wastes from households are contained in green bins and are brought to the compost plant. When it arrives, it is sorted out again to take out the plastics, glass and metals (Tcpm, personal communication, June 15, 2010).

Afterwards, thick matter is crushed to reduce its size until it is less than 140 mm, and then it is taken to one of the tunnels where compost is mixed and processed aerobically with the help of computers, which control the temperature, moisture and oxygen to ensure that bacteria, fungi and microorganisms do their job. The resulting ash from incineration is also included. During this stage of pre-compost, temperature must remain at 60 °C, to remove weeds. During the process, biofilters containing bacteria and fungi purify it. After 12 days, most of the material is composted and taken to an area for the post-compost stage, where it stays for a week. The final result is compost which can be used as potting soil in agriculture.

In 2009, over 92,000 tons of waste was recovered for compost elaboration, which is almost 7% more than in 2008. This is due to increased efficiency of the composting plant: the trituration of thick matter improves the process. After composting through tunnels, waste is filtered through a strainer, where wood and other rubbish from the compost are collected to be used as a biofuel for producing energy, steam heat or electricity. Of the 147,894 tons of biomass burned in 2009 at the incinerator BEC, 20% of it came from waste in the composting plant (Twence Annual Report 2009).

The manager of the composting process in Twence (mcpT), mentioned that each year, they try to be greener with the waste management by producing electricity, biomass and heat from the incineration of waste. He also said that one of the barriers encountered is that they are paid by ton of waste, which has decreased from 50 Euros per ton to 25 Euros per ton, and now they have to look for alternatives to be profitable. Another barrier is the smell that emerges from Twence, which generates complains from the neighbours. The compost plant is forced to prevent this by cleaning the air before it goes back to the atmosphere, and it costs a lot of money. The mcpT also said that they earn the money from the organic waste that they receive, but they must pay farmers to get the compost produced. Another barrier mentioned is that the plant depends on what the government decides on waste legislation.

An area of opportunity is that soil in the Netherlands is getting (mcpT, personal communication, June 23, 2010), so it is expected that people will use more compost from the Twence plant as a soil improver. Finally, the mcpT believes that the tours in the Twence plant are of vital importance because if people know about this type of projects, then they are motivated to cooperate in separating their waste and become potential customers of the products that come out of their waste, enabling a circular economy model for sustainability.

This research addresses the Dutch composting process, which is performed in an industrial scale, but considers the legal and economic instruments applied to stimulate citizens and in particular householders to sort out their wastes. The Dutch case's approach for the implementation of this waste management policy is top-bottom, which can still be used as a reference for future recommendations in the Mexican situation.

12.3.4 Mexican Case

In this section, the Mexican case is studied considering the municipal landfill scheme as a way to manage residues, as well as the particular case of the community compost model used in a residential area in the same municipality (Atizapán de Zaragoza).

12.3.4.1 Compost Plant in *Puerto de Chivos, Atizapán de Zaragoza*

The municipality of Atizapán de Zaragoza participates in a programme of municipal compost which started in July, 1999. Its objective is to reduce the quantity of organic waste that is disposed into the landfill *Puerto de Chivos*, in order to decrease the polluting effects and extend the life of the landfill (Villegas and Franco 2010). According to statistics, if every organic residue was composted, the waste disposed would be reduced by 49%.

The programme consists in the participation of 43 residential areas which separate their waste, but 110 residential areas do not participate due to lack of knowledge and dissemination of the programme in the municipality, which is a barrier for the continuity.

The perception of the Mexican compost manager (Mcm) of the citizens' behaviour towards waste separation is that they believe that even if they separate their waste, it would be mixed again when it arrives to the landfill.

Although they receive visits from schools and international and national organisations that are interested in the compost programme, it is vital to have more dissemination of the benefits. Another obstacle is that this programme's continuity depends on political cycles, which means that every 3 years, when elections take place, the personnel changes. This affects the development of the programme and makes it difficult to improve it when there is a great deal of staff turnover.

In terms of the compost production process, the bio-inputs used are the organic residues from these households: leaves, grass and splinters from parks and green areas in the municipality and manure from cows and horses from ranches nearby and some unidentified restaurants. The landfill receives 500 tons of waste per day, but can only recover 0.8%, which means that it produces 54 m^3 of compost per month (INE 2010).

When the organic waste arrives to *Puerto de Chivos*, scavengers do a manual separation of the garbage because it arrives in plastic bags, which are removed in order to make piles of the residuals found in them. As for the pruning material and grass clippings, the trituration process is done and then mixed in an open space. The production of the organic fertiliser is an aerobic process done in 1000 m^2 of land. The machinery and tools used for producing compost are two backhoes, shovels, a sieve and two big shredders. The compost is only covered with plastic when it rains, and the rest of the waste that could not be used for compost is disposed into the landfill (Mcm, personal communication, June 2010).

Bio-inputs are mixed and then placed among alternating layers of grass until the temperature reaches 70 °C. The process takes 3 months (besides the time of construction of the pile), and it is watered every 3 days, depending on the season, with treated wastewater going through a pipe. The backhoe flips are made once a week or when machines are available.

The potential hydrogen is not measured. They only have an approximation from the carbon-nitrogen relation, by using "brown" inputs like fallen leaves and manure, which represent the carbon, and the "green" inputs like organic material, and grass, which represent the nitrogen. For the aeration of the compost, they installed tubes. The compost is not sold, just given away to use it in parks or at the request of citizens. The quality is not measured through chemical analysis, even though there is a norm in Mexico City that establishes the standard.

In the economic aspect, *Puerto de Chivos* does not have a long-term plan for the compost plant. In terms of budget, the costs of the plant per year are $201,896 Mexican pesos[1] (INE 2010). The feasibility of the plant is not known because they do not make annual reports of their performance; they have a lack of transparency in their management, and the municipality does not sell the compost, so there is no return of investment.

[1] Equivalent to $16,023 USD (exchange rate: $12.6 pesos per USD in July 2010) according to http://www.banxico.org.mx/portal-mercado-cambiario/

Although there have been some efforts in Mexico to implement composting plants, they have been an economic failure and in most cases have been closed. Among many reasons that have led to discard the use of composting in Mexico is the lack of market for this product, which use is restricted to agriculture as a soil improver (Semple and Fermor 2001).

On an exploratory visit, we had an interview with the manager of the municipal compost plant. We asked him about the production process of the organic fertiliser done in *Puerto de Chivos*, and he answered:

> The process could take from 3 to 5 months, depending on the level of humidity and temperature, and as a result 70 m^2 of compost are produced.

Atizapán de Zaragoza is biologically treating the waste, but it does not have the necessary technology to do it. We asked the compost manager what would they do if more people separated their waste: would they be able to increase their production capacity? He answered the following:

> Yes, we have enough space to produce compost… If more people separate their waste in their houses, it is better for us because although it implies more work, it is worth it.

Since this programme has some deficiencies, a residential area called *Bellavista*, which is one of the residential areas that participates in the programme of waste separation, started a community compost plant. From the research done in the community compost plants in the United Kingdom, the variable of social participation was identified as the most important variable because it can start and drive a community compost plant or become a significant barrier. That is why it is important to know what the drivers and barriers for the Bellavista's community compost plant are.

12.3.4.2 Study Case: Club de Golf Bellavista

Club de Golf Bellavista is a residential area located in *Atizapán de Zaragoza*. It has 80 hectares and 345 houses. The total households' waste had been taken to Atizapan's landfill called *Puerto de Chivos* before the local compost plan started in 2008.

This initiative started with one neighbour, who is an environmentalist, when she proposed the idea of a compost plant inside the residential area. She convinced the neighbours' associations, which then realised that the problem of waste management in Mexico City is not solved by local governments yet. They were motivated to use less pesticides, chemical fertilisers, ammonium sulphate and urea in green areas and also realised they could take advantage of the organic residues instead of sending them to landfill while having economic benefits.

Subsequently, the programme was announced to each neighbour, but only a small percentage of them separated their waste. It is also important to mention that this residential area has a social club with a golf course, a restaurant and a kinder garden. Most of the food residues from the restaurant are used in the elaboration of compost; gardening waste from households is used as well, but it is not enough because if more people separate their waste, the compost produced could be used not only in green areas and some of the houses' gardens but in the golf course, thus producing an even greater impact (Martinez 2009).

It is important to mention that the green residues from pruning the golf course most of the times are not used for the elaboration of community compost. However, this would have to be done if the official norm NOM 140 SEMARNAT 2005, which establishes the general environmental requirements for golf courses and real estate development that includes golf courses, is approved. This norm established that the grass and other organic compound derived from the maintenance of the project areas are meant to be used for compost, fodder or other purposes to ensure reuse (SEMARNAT 2005). This norm is not mandatory yet.

According to the manager of the compost plant in Bellavista (mcpB), they could represent a sustainable model for waste management in a community scale:

> We have an ecological corridor, which can be visited by the Bellavista neighbors and general public. We produce compost and we also have a nursery area, compost produced by worms and an area of medicinal herbs. This is a scale model of what people can do to recycle organic waste. The next step is to increase the area destined for the compost area in order to sell at a low cost the compost produced here.

The area that they have is an ecological corridor which has medicinal plants, a plant nursery, and also vermicompost. This could also serve as a model for other residential areas with similar characteristics as Bellavista.

The continuity of the community compost plant in Bellavista relies in the fact that there is an institutional figure in charge of the production of the organic fertiliser called *Asociación de Colonos de Bellavista*. But it is somehow informal in its operation due to lack of official records on waste received at the compost plant. The area destined to treat organic waste is approximately 75 m^2, in an open space, and they have to be very careful with the right production of compost in order to prevent the presence of vermin and odours.

The advantage is that the quality of the compost fulfils the standard that Mexican regulation states for good soil improvers. Since the technique to produce the fertiliser is accelerated by the bio-inputs, this could represent a success factor.

Club de Golf Bellavista also considered different alternatives in order to stop using pesticides, but the most feasible was recycling organic waste through aerobic biological treatment, commonly known as compost. Economic, social and environmental aspects were considered when the decision was made. This treatment is not expensive to implement in residential areas, as explained further by the mcpB.

In order to start modelling the composting plant in Bellavista, the scheme had to be divided into levels by using the diagrams of Forrester (Forrester 1991), which also constitute different phases: compost in process, householders who separate their waste, required personnel and maintenance costs.

Therefore, translated into mental models, the behaviour of each level can be described by using the following:

- *Compost in process:* steady increase, reaching a peak due to the limited space for composting and other resources.
- *Householders who must separate their waste:* it is contemplated as a constant increase, where householders start to separate their waste until it reaches its peak, which is 345 households.
- *Required personnel:* it is contemplated that the number of people in charge of the composting process increases when the quantity of inputs is more, but since the space is limited, the maximum level of people engaged in the process is four.
- *Maintenance costs:* even though there are some fixed costs such as workers' wages, and gasoline for the garbage truck, the model reflects that the maintenance costs are reduced (Forrester and Martin 1997).

For instance, the quantity of *organic matter* is a rate fed by the quantities needed to produce compost. This rate is linked to the level called *compost in process*. After that, the level *compost in process is* translated into a rate called *resulting compost*. The *resulting compost* is the result of all those variables. This is the way the model behaves.

It is a complex model because it involves many variables, but it has been simplified. First, the time frame within which the scheme is assessed lasts 24 months, which implies that many variables are remaining constant, like the production area, the machinery and tools used, an approximation of the quantity of waste produced in each household per month and the number of households. The evolution of the process is analysed every month during this time frame.

It is important to know that, in order to work, this particular scheme has to consider variables like an area destined to process compost and people trained in waste management through composting. Management of the compost plant is mainly voluntary through the neighbours' association of *Club de Golf Bellavista*, where the only ones paid are the waste collectors which are trained in order to know where the organic residues have to be taken (mcpB, personal communication, June 2010).

Throughout the year, pruning of house gardens, green areas and the golf course is done. This translates into raw materials for the production of compost, which is then used as fertiliser in the green areas mentioned. Another aspect is the separation of garbage from the neighbours in organic and inorganic. This makes it easier to use organic waste to nourish the compost, so the more it increases, the more fertiliser for landscaping there is.

There are some periods in which more organic waste is produced, such as winter, when Christmas trees are discarded; those periods demand more personnel for the compost production (mcpB, personal communication, June 2010). In the model, it is assumed that a person can handle up until 3000 kg of organic matter. When the personnel reach four people, it is considered that they are enough since the space of the compost plant is relatively small, and in this way, the maintenance costs for the green areas do not increase dramatically.

Organic matter produced in Bellavista is later processed and packed into bags that weight 30 kg each. As a result, in winter more compost is produced and exceeds the compost demand for the maintenance of green areas in the residential area, which is why it can be sold.

The interaction of different "systems", mainly the economic and social ones, is considered in order to prove that this could become a sustainable model for waste management in residential areas with similar characteristics. The generation of compost could save money because the organic material is used in the maintenance of green areas; it reduces costs when the amount of trash collected decreases and fewer trips are made to the landfill. This implies lower costs in waste collection and disposition of residues into the landfill.

The level of maintenance costs decreases each month until it reaches the lowest point on the 8 month of the first year. The economic benefits of this compost plant include the creation of three more jobs, sales of the compost produced and savings, so it is a feasible model. The typical maintenance costs before the implementation of the composting programme were approximately $55,212.00 MXN pesos[2] per month (mcpB, personal communication, June 2010), but when the compost plant started, the typical costs were discarded and replaced with the organic residues as bio-inputs for compost production.

The model represents the behaviour of the programme if all organic residues were used in the production of organic fertiliser. The assumption is that the number of neighbours in *Bellavista* who separate their waste and used for composting increases. Savings are presented although trained staff has been hired for the operation of the plant.

The total maintenance costs after 2 years of operations are $138,530.42 MXN pesos,[3] which imply savings of $45,000 MXN pesos[4]; therefore, it is a viable proposal to be implemented in other neighbourhoods without additional costs. It is important to mention that without the neighbour's participation, this scheme is not possible.

In the proposal, when the resulting compost exceeds the compost required for the green areas, bags of compost could be sold to external members of the community in $90 MXN pesos[5] each. As a subsequent proposal, the compost could also be used in nurseries. In a period of 24 months, the compost sales could generate an income of $305,211.60 Mexican pesos.[6] In the long run, vegetables and plants derived from the *Bellavista* nurseries could be marketed. Therefore, it is a sustainable project as it provides economic, social and environmental benefits.

It is important to note that social participation in the separation of organic waste could represent a barrier for the success of this project. In order for it to work, it is vital to inform that the only bio-inputs used for the production of compost from kitchen are vegetables, fruits, eggshells, grass, branches and Christmas trees.

[2] Equivalent to $4381.9 USD (exchange rate: $12.6 pesos per USD in July 2010)

[3] Equivalent to $10,994.47 USD (exchange rate: $12.6 pesos per USD in July 2010)

[4] Equivalent to $3571.42 USD (exchange rate: $12.6 pesos per USD in July 2010)

[5] Equivalent to $7.14 USD (exchange rate: $12.6 pesos per USD in July 2010)

[6] Equivalent to $24,223.14 USD (exchange rate: $12.6 pesos per USD in July 2010)

The compost plant manager (mcpB), explained that not every household separates its garbage; approximately 10% of the neighbours have not participated ever since the compost plan was opened in 2008. Because the space to produce compost is small, it does not seem as a priority to promote more waste separation at the time, but the space is expected to increase, and generating more awareness among the Bellavista's neighbours would then be necessary.

Since social participation could constitute an important barrier for the continuation and growth of this community compost plant, this led to suggestions to increase the awareness of the environmental and health benefits of this. One of the main suggestions was to inform the community about the compost plant inside the residential area and to inform the neighbours what kind of organic waste could be used in the production of the compost or organic fertiliser. This is an important aspect because they assume that all organic waste can be used for the production of compost, when in reality some bio-inputs are not used, such as meat and bones.

The following results represent the most relevant variables analysed in the questionnaire on social participation in Bellavista, which influence directly on the neighbours' attitudes and behaviour towards waste separation. The results are the following: 40% of the neighbours who know about the compost plant separate their waste; 33% who know about the compost plant do not separate their waste. The main reason that keeps them from separating waste is that neighbours think it is useless to do it because it usually is mixed once it is picked up. This highlights the lack of a good communication between the compost plant managers and neighbours. Neighbours should know that if they separate their waste, the organic residues could be used in the elaboration of compost used in green areas, and they could also request it for their gardens.

After knowing this situation, it was important to know if the behaviour and attitude towards waste separation is related to gender, age and education level. The majority of the respondents were women between 45 and 65 years (73.3%), of which, only 33.3% separate their waste. When considering the education level of the neighbours, the majority studied until college (56.7%), but when it comes to separating their waste, 64.7% of them do not separate their waste. This does not seem logical, that is why people that do not separate their waste were asked if they had a particular inconvenient that prevents them from separating their waste: 29.4% said that they don't have the habit of doing it or they may forget, and 64.7% think it is useless because waste is mixed anyway.

In this survey, the subjective norm, which considers the influence of family and friends in a person's behaviour towards the separation of their residues, is measured (Taylor and Todd 1997; Vicente and Reis 2008). The Bellavista community is not necessarily influenced, 36.7% said that they have never talked about their waste management habits, and the people who have, say that only 26.6% said that if a family member is environmentally conscious and active, then they would start to separate their waste.

It is relevant to mention that 30% of the respondents do not spend money on fertilisers and are not willing to contribute economically with the compost plant in

Bellavista, and 23.3% spend more than $400 pesos[7] in fertilisers per year, but when they were told that they could receive free compost and asked if they were willing to pay an amount per year, only 13.4% said they would pay from $1 to $300 MXN[8] pesos per year.

It was also very important to know to what degree some mechanisms could convince the neighbours to participate in the separation of their waste. The *Bellavista* neighbours pointed out that visiting the compost plant, getting information about the benefits of using compost and getting free compost are the best mechanisms to convince people to participate. On the other hand, the mechanisms that would not convince them at all were to participate in a competition of the household that separated more waste and attend to workshops about waste management.

The last question was an open one in order to know the opinion of the neighbours when it comes to suggestions to encourage community participation in *Bellavista*. One of the main suggestions was to inform through the neighbour's newsletter (56.7%) and putting different colour of trash cans outside the houses in order to make easier the organic waste collection (16.7%).

12.4 Comparative Study

The purpose of comparing the three schemes is to get to know the conditions, drivers and barriers of the compost plant created through community participation and specifically in residential areas. These analyses are in Tables 12.1 and 12.2.

Table 12.1 summarises the answers of compost plant managers concerning the drivers that led to the creation of their existing compost plant. This is relevant because although the scales of the composting programmes are diverse, the drivers are similar. In the Netherlands, it is prohibited to use landfills; the only exception is when the production capacity for incineration is exceeded. Most of the cases have in common that the programme started in order to reduce the amount of waste sent to landfills; only in the case of *Bellavista*, the driver was to stop using chemical fertilisers in the maintenance of the green areas.

The second driver that stands out is producing compost instead of buying it. This has the benefit of saving money, so it is considered an advantage. In the United Kingdom, they consider it a main driver because they explained that during the World Wars, they did not have food, so they had to grow their own food, and with the organic residues, they produced compost.

The third driver is to reduce the use of chemical fertilisers, which is the main driver for the Bellavista case, but not for the other three cases, although the compost produced in Twence is certified as a soil improver, and it is considered that the use of chemical products just damages the soil.

[7] Equivalent to $31.74 USD (exchange rate: $12.6 pesos per USD in July 2010)

[8] Equivalent to $23.8 USD (exchange rate: $12.6 pesos per USD in July 2010)

Table 12.1 Analysis of drivers to implement community compost plants (Own contribution)

Compost plant	Reduce waste in landfill	Legal incentives	Main reason to reduce the use of chemical fertilisers	Composting as an alternative to buying new compost
Case 1: Residential Area Club de Golf Bellavista, State of Mexico	Not a main objective or motivation	No legal incentives or fines	Main motivation	Not a main driver, but it is considered an advantage
Case 2: Puerto de Chivos, Atizapan de Zaragoza, State of Mexico	Main objective of the programme, although there are no legal sanctions for land filling waste	No legal incentives or fines	It is not considered a driver	Not a driver, just a consequence of doing the compost, although it is not enough for all the green areas in the municipality
Case 3: United Kingdom (Community Network)*Greenway Consulting *Brighton Community *Lower Slaughter *Cwm Harry and Trust*Denton Parish Council	The UK measures how much waste was kept from going to landfills	No legal incentives to do community compost. Just agreements with local authorities	They do not consider this a driver for the community to participate	They consider this a main driver
Case 4: Twence compost plant in Hengelo, Netherlands	Strict regulations Tax payment to use landfill	Yes, legal incentives present	They consider that the compost is very important in order to improve the soil, whereas the chemical fertilisers just damage it	The compost plant from Twence even pays farmers to get their compost

The legal incentives do not work in all cases, only in the Netherlands. Since there is no mechanism to encourage the creation and continuity of composting programmes, it becomes difficult to compel the performance of these programmes. The Netherlands fines the householders that do not separate their waste.

Once the drivers are known, it is vital to analyse the necessary conditions in which a composting programme works. The conditions could also represent a barrier for the implementation of such a programme. Table 12.2 analyses the managers' answers.

The first condition is social participation; this implies participation of the community by separating their waste and, in the case of the United Kingdom, volunteers to process the waste and the management of the compost plant. This could represent a barrier because it would be very difficult to process the waste if it is mixed with

Table 12.2 Analysis of conditions/barriers to successfully implement community compost plants

Compost plant	Social participation	Economically feasible	Infrastructure to do compost	Quality of compost under regulation	Environmental impact of the project
Case 1: Residential Area Club de Golf Bellavista, State of Mexico	The participation is low, but so its production capacity	This scheme is economically feasible, because in time compost this could be sold if there were more bio-inputs involved	The area destined for the production of compost is 75 square meters approximately; they have 2 shredders and 3 shovels	The norm NTEA-006-SMA-2006 states the compost quality standards, in order to sell it	They do not measure this in a formal way
Case 2: Puerto de Chivos, Atizapan de Zaragoza, State of Mexico	The participation is low, only 28% of the residential areas in Atizapan de Zaragoza are included in the programme, from which only 64% separate their waste	Not feasible; it is more expensive to produce compost than to landfill all the waste	They have enough space to grow in production capacity, but they currently lack infrastructure to process all the organic waste from the municipality	The norm NTEA-006-SMA-2006 states the compost quality standards, in order to sell it, but they do not do chemical analysis of the compost produced	There is no measurement or report of the environmental impact of the compost plant 'Puerto de Chivos'
Case 3: United Kingdom (Community Network) *Greenway Consulting Scotland. *Brighton Community Compost Centre. *Lower Slaughter Community Composting. *Cwm Harry and Trust *Denton Parish Council	Important but not vital because they use other green waste from pruning	It is important to be feasible, but not a requirement because in some cases they get paid to treat the organic residues instead because they avoid to send it to the landfill	Each community compost has the space necessary to produce compost, and the proper machinery to do it, close from the community	The norm PAS 100:2005 states that the specification of composted materials in order to be able to sell it. If they do not have a certification, the compost will be considered waste	They do not measure this in a formal way. They quantify how much organic waste they receive, and that it is kept from going to the landfills

(continued)

Table 12.2 (continued)

Compost plant	Social participation	Economically feasible	Infrastructure to do compost	Quality of compost under regulation	Environmental impact of the project
Case 4: Twence compost plant in Hengelo, Netherlands	Vital. The compost cannot be produced if people do not separate their garbage	The process itself is feasible because they process waste through an industrial scale and they get paid to process organic waste, but they cannot sell it yet	The infrastructure used in Twence was created to have the capacity to produce compost with organic residues of the Twence region	They have certifications like BRL Keurcompost: branche eigen certificaat compost and BGK: voor de afzet van compost in Duitsland	They measure their gas elaborate and make annual reports

inorganic residues and the costs could increase to separate every single organic residue.

The economic feasibility is an important aspect because, if it is not profitable, the programme has to stop. Although the case of *Atizapán de Zaragoza* is not profitable, they have to produce the compost because it is an obligatory programme.

The third condition is to have the proper infrastructure to process all the organic waste, and, in every case, it is considered as a must.

The next condition is that the compost should have quality standards stipulated in the regulation of each country. In the case of *Bellavista* and *Atizapán de Zaragoza*, there is a norm that states that in order to sell the compost produced, it has to be analysed and satisfy the chemical characteristics. *Bellavista* has analysed its compost once, in April 2010, but the government of Atizapán de Zaragoza, who handles the compost plant, has not done any chemical analysis. This could represent a barrier in order to market the compost. The Twence and United Kingdom cases also have to certify their compost because if not, it is considered waste.

The fifth condition is to have an assessment on the environmental impact of each programme. Most of the cases do not have a formal report on their environmental impact. The only assumption is that it is positive because less waste is taken to the landfill. In the case of Twence, each year they present a report on the gas emissions of the composting process because it is a legal requirement to do so.

These conditions are taken into account in order to transfer the community compost scheme to other residential areas in Mexico, starting with the municipality *Atizapán de Zaragoza*, and build a composting network. The direct benefits go to the community: they receive compost for their gardens, and it represents savings for the neighbours' association, as seen in the Bellavista case.

12.5 Conclusions and Recommendations

As seen before, there are two different international schemes that could contribute to suggest how to handle Mexican waste management. The Netherlands presents an top-down perspective and the United Kingdom a bottom-up perspective. It can be said that Mexico is not yet prepared to adopt Dutch approach. Even further, it is rather proposed to analyse a bottom-up approach that starts in a community compost plant in *Atizapán de Zaragoza*, where the organic residues are taken to be handled, and the rest are taken to the landfill *Puerto de Chivos*.

As a result of the comparative analysis in Sect. 12.4, it can be concluded that there are some similarities among the countries of the Netherlands, Mexico and the United Kingdom in some of the legal, social, economic, environmental and operative aspects, which are recognised in this research and mentioned in order to point out if they represent a driver or a barrier for the implementation of community compost plants.

The first aspect is the legal framework, which was addressed through the additional research question "What are the existing legal frameworks in all three countries that permit this sort of small scale projects?" The response is that in Mexico and the United Kingdom, a small-scale composting programme is not regulated nor prohibited; meanwhile the local government's programme for the composting of organic residues does not work correctly because of the lack of dissemination and enforcement in the case of Mexico; and in the United Kingdom, this way of treatment is not allowed because some organisations consider that it is not sanitary to make compost out of animal byproducts. As a result, this research suggests the creation of community composting programmes, this based on the analysis of both cases.

In the Dutch situation, there is a legal framework that establishes every residue should be reused, recycled and treated before it goes to the landfill. The programmes work because householders separate their waste, otherwise they would obtain a fine. In Mexico and the United Kingdom, at the time this research was carried out, there were no incentives (positive or negative) to start segregating waste from householders.

An aside research question is related to the social aspect of stakeholders. In Mexico, the main stakeholders are the neighbours living near the compost plant, whose main driver was to use less chemical fertilisers and therefore decrease the damage to human health. It is important to know that for the continuity of this programme, there is a neighbours' association, which manages the collection and waste treatment.

In the United Kingdom's cases, volunteers are in charge of producing compost, but there is always a group of citizens who manage the project in order to ensure the success and permanence of these programmes.

In the Netherlands' case, there is a combination of public and private stakeholders because waste collection is done by the local government, but waste treatment is done by a private company. The investment is both public and private, which ensures the permanence of the programme. It is important to mention that due to the composting system size, the comparison between the Dutch system and the two others of this research was not possible.

It is also important to consider that with the purpose of implementing a successful compost plant like Twente, citizens' participation by separating their waste is vital, and the employees must be trained. All the stakeholders have an important role, but when it comes to the implementation of a composting programme, the management is the key to give permanency to such programmes. And such lesson can be extended somehow to the communitarian composting plants in Mexico and the United Kingdom.

In order to respond the research question mentioned in the abstract, an additional question was formulated: "What were the main actors' motivations and incentives to engage the composting plant in the community in order to process organic waste?" The main motivations found in all the case studies were that by engaging in composting organic waste, less quantity of residues is sent to the landfills, and less money is spent in buying fertilisers, because the compost produced can be used as a natural fertiliser, and it also can be sold when its quality is proved.

The question to address the social, economic, environmental and operational variables was formulated as "What are the differences and similarities in terms of standards for the socio-economical-political-cultural contexts among the three countries?"

The role of political actors in public administration does not influence the waste management schemes of the Netherlands and United Kingdom, since Twence (NL) is a combined system (public and private) governed by market forces. While in the UK waste management is done through community organisations, they have independence and continuity despite changes of government. This situation is not given in the Mexican context, as landfills and local compost plants' management is centralised and managed by municipal governments. This difference in management may represent an incentive in the implementation of community composting, which would give them autonomy and continuity to the project.

The variables considered for the analysis of conditions for the successful implementation of a compost plant were social participation, economic feasibility, infrastructure to do compost, quality of compost under the regulation and the environmental impact of the project. These could constitute a condition or a barrier that could intervene in the success of a composting programme.

The first variable is the social participation; this is vital because if the community does not separate their waste, the raw material for the elaboration of the compost could be polluted with other sorts of inorganic material, and as a consequence, the quality of the compost diminishes. It is also important to mention that in Mexico, there is an issue concerning the environmental culture because there is a perception that even if citizens separate their waste, it would be mixed again, which is why the composting programmes do not work.

The second variable is the feasibility of the programme. Most programmes sell the compost produced (UK and NL) or make agreements with the local government to get paid for treating the organic waste instead of sending it to landfills (UK and

NL), and in the Mexican cases (Puerto de Chivos and Bellavista), they use the compost in green areas instead of fertilisers, which produces savings. The fact that the compost produced can be sold and the situations mentioned above makes a compost plant feasible.

The third variable is the infrastructure to produce compost; the minimum requirements are a space to produce compost and store raw materials, shredders and shovels. The quality of the compost is also measured in each case, and it is considered an aspect that is important in order to sell it. The last variable is the environmental impact of the project, which is not evaluated in small-scale waste management, only in Twence, because it is legally established in Netherlands, although it is important to propose this evaluation in each case studies.

The final question is related to the operational aspect: "How does the project management contribute to the permanency of community composting plants?" The main answer was through formalising the organisation that operates the community compost plant, with clear roles and responsibilities of each member who participates in the management of the community compost.

And finally the success aspects of the existing compost plants and how they can be used as a reference to improve the compost plant in Bellavista are hereby enlisted in the form of suggestions to improve. the current residential scale model in Club de Golf Bellavista:

- In order to sell the compost produced, it has to be a high quality one, so it is important to have the compost analysed in its chemical characteristics regularly and getting an international certification to prove the quality of it as a soil improver.
- Start an awareness campaign in order to promote the compost plant in Bellavista and invite neighbours to visit the facilities.
- Elaborate annual reports about the organic waste received and processed and the resulting compost, including the economic performance.
- Organise trainings for compost plant operators.
- Approach the local government for technical and economic support.
- Inform the neighbours about the compost benefits and the way they could get free compost for their gardens through newsletters and the installations of trash cans of different colours to collect the organic waste more efficiently and start workshops.

To sum up, it can be concluded that for a community compost plant to work, there should be a dissemination of the programme. To be profitable, the composting process needs to ensure the quality of the final product in order to sell it, have a suitable production area to prevent odours and the presence of pests or vermin, as well as institutionalising the activities by means of manuals of good practices in the management of organic waste to produce compost and annual reports on its performance. With these conditions, the community compost scheme could translate to other residential areas with similar characteristics.

As a next step and a recommendation, a community compost network could be created as a support for those who start a community compost plant to share recycling best practices and technical and economic advice.

In this regard, as part of a pilot programme, the composting network could be implemented in the municipality of *Atizapán de Zaragoza*, with 153 residential areas, where there are neighbourhood associations. By doing this, 1046.52 tons per year of organic waste would not be taken to the landfill *Puerto de Chivos*. Besides, 469 direct jobs are expected to be created for the operation of the compost plants. This could also lead to an agreement with municipal authorities for the donation of land for the waste treatment process.

When the barriers of locally composting systems are overcome, the collection and treatment of USW can be replaced by one industrial composting system. Although this perspective is not currently feasible for Mexico, it is crucial to support governments to consider a more sustainable way of dealing with household waste in the long term.

Finally, some recommendations for strengthening this research are to have a more representative sample of the population and to expand the number of cases that have a similar waste management as Bellavista.

References

Álvarez AE (2002) Historia de la basura en la ciudad de México. http://www.geocities.com/camp_pro_amb/historia.htm. Accessed June 2010

Association for Organic Recycling (2010) http://www.organics-recycling.org.uk. Accessed July 2010

Bernache Pérez G (2006) Cuando la basura nos alcance: el impacto de la degradación ambiental. Centro de Investigaciones y Estudios Superiores en Antropología Social, Distrito Federal

Business and Science Park (2010) http://www.sciencepark.nl/en/location/. Accessed 22 July 2010

Castillo Berthier HG (2002) CIENCIAS 20

Community Compost (2010) http://www.communitycompost.org. Accessed July 2010

Compost Network (2008) http://www.compostnetwork.info/index.php?id=40. Accessed July 2010

Compost-it.org.uk (2010) www.compost-it.org.uk. Accessed July 2010

Cylch Community Recycling Network (2010) http://www.cylch.org. Accessed July 2010

Dennis N, Baxter A, Darwin N (2006) Potential health risks to humans from birds, mammals and insects associated with UK waste management operations: A literature review. Bird Management Unit, United Kingdom

Forrester Germeshausen JW (1991) System dynamics and the lessons of 35 years. Massachusetts Institute of Technology, Cambridge, MA

Forrester JW, Martin LA (1997) Beginner modeling exercises. MIT System Dynamics in Education Project

Greenway Consulting (2010) http://www.greenwayconsulting.co.uk. Accessed July 2010

Greenwichmeantime (2010) http://www.greenwichmeantime.com/time-zone/europe/uk/map.htm. Accessed October 2010

Herrera Massieu R (2004) "Cómo y por qué separar la basura. Una solución al problema de los residuos sólidos en la Ciudad de México. Distrito Federal: Servicios Industriales Peñoles S.A. de C.V. Secretaria de Medio Ambiente del Distrito Federal

Instituto Nacional de Ecología (2010) http://www2.ine.gob.mx/publicaciones/libros/499/experiencias.html. Accessed June 2010

La república (September de 2008) Se acaba el tiempo para la basura de la ciudad de México

Martinez M (2 de October de 2009) Questionnaire on Compost plant experience with community participation. (V. Plasencia Vélez, Entrevistador)

Medellin C (2015) Gestionan ampliar 15 anos mas el relleno de Puerto de Chivos. Revista Mexiquense. http://www.reportemexiquense.com/infomacion-general/gestionan-ampliar-15-anos-mas-el-relleno-de-puerto-de-chivos. Accessed July 2010

Mejia F (Mayo 2016) La basura genera 16% de la contaminación. Milenio

Ministry of Housing, Spatial planning and the Environment (2001) Legislation, waste in Netherlands. Netherlands: Directorate-General For Environment Protection

Organics recycling (2010) http://www.organics-recycling.org.uk. Accessed July 2010

Restrepo, Phillips (1991) La basura, consumo y desperdicio en el Distrito Federal. Centro de Eco desarrollo, México

Secretaria de Obras Distrito Federal (2010) http://www.obras.df.gob.mx/servicios_urbanos/residuos/rec_trans_sel_final.html. Accessed Sept 2010

Semple KT, Fermor TR (2001) Impact of composting strategies on the treatment of soils contaminated with organic pollutants. Environ Pollut 112:269–283

Slater R, Davies P, Gilbert J (2004) The state of composting in the UK. The Composting Association, United Kingdom

Taylor S, Todd P (1997) Understanding the determinants of consumer composting behaviour. J Appl Soc Psychol

Twence (2010) http://www.twence.nl/en/over% 20twence Accessed June 2010

Universidad Autonoma de Guadalajara (2010) http://www.uag.mx/27/genetica/gpos.htm. Accessed Sept 2010

Vicente P, Reis E (2008) Factors influencing households' participation in recycling. UNIDE, Department of Quantitative Methods, 140–146

Villegas A, Franco R (2010) Exploratory Questionnaire to responsible of composting process in Puerto de Chivos, Atizapan de Zaragoza. (V. Plasencia, Entrevistador)

VROM (2010) Recovered in July de 2010

Waste Online (2005) http://www.wasteonline.org.uk/resources/InformationSheets/Compost.htm. Accessed June 2010

Chapter 13
The Future of Circular Economy and Zero Waste

María-Laura Franco-García, Jorge Carlos Carpio-Aguilar, and Hans Bressers

Abstract This chapter provides a reflective analysis of the contributions focusing on how the chapters contribute to circular economy in two main themes that will define the future of circular economy research: challenges and opportunities and effective collaboration with stakeholders. Lessons learned and commonalities are also discussed to finally draw some conclusions and recommendations to further elaborate the research agenda of this field.

Keywords Circular economy · Collaboration with stakeholders · Zero waste

13.1 Country-Specific Challenges and Opportunities

One of the lessons from the chapters is that challenges and opportunities have been so far somewhat generically described in the literature, which tends to assume a universal view of challenges and opportunities, where the lessons from developed countries can be without a problem translated to developing countries. We argue that a more contingent approach is needed, taking into account deep structural, institutional and cultural differences between countries. In other words, our understanding of circular economy (CE) challenges and opportunities needs to be geographically and socially embedded to understand country-specific differences. Accordingly, in Chap. 2, Dieleman and Martínez analysed opportunities and challenges of CE in Mexico from a wide systemic perspective inspired by the concept of national innovation systems and its emphasis on distinctiveness of contextual conditions. To frame their discussion, they used the components of such concept, i.e. (i) "market trends and conditions"; (ii) "competitiveness and productivity"; (iii)

M.-L. Franco-García (✉) · H. Bressers
University of Twente, Enschede, The Netherlands
e-mail: m.l.francogarcia@utwente.nl; j.t.a.bressers@utwente.nl

J. C. Carpio-Aguilar
Walmart Mexico, CDMX, Mexico
e-mail: jorge.carpio@walmart.com

M.-L. Franco-García et al. (eds.), *Towards Zero Waste*, Greening of Industry Networks Studies 6, https://doi.org/10.1007/978-3-319-92931-6_13

"the political and regulatory framework"; (iv) "education, training and knowledge transfer"; and (v) "learning culture". As part of the opportunities, (i) the market conditions and long-term market trends in the country are positive; (iii) the regulatory framework through the country's National Development Plan aims at stimulating green growth; (iv) since the 1990s, environmental education nationally exists as well as knowledge transfer systems; (v) Mexicans' openness to new technology induced practices. In contract, serious challenges are also mentioned: (i) markets are not flexible, and there is lack of bank credits allocation in the long term; (ii) a culture that restricts innovation and creativity and a desire to shape and control the future leaves Mexico with poor seizing opportunities; (iii–iv) discrepancy between the creation and execution of policy and the environmental education plan is another real challenge; (v) the needed rather experimental and learning-by-doing educational approach to address CE shows limitations of the current Mexican learning culture. The authors emphasised this last element because it influences how the economy and society function and therefore represents one of the biggest challenges to overcome.

In particular within the waste sector, in Chap. 3, Ghinea and Gavrilescu analysed the Romanian waste management policy framework and practices to identify challenges and opportunities of CE goals. As a result, they suggested strategic solutions for integrated waste management and policies for its improvement. With the help of the life cycle assessment methodology, eight waste management scenarios were assessed. The first one was used as baseline (collection, transportation and landfilling) for the comparison with seven hypothetical scenarios which are constructed following the principles of Lansink's ladder (2010). As mentioned in Chap. 1, Lansink's ladder emphasises the relevance of preventing waste generation during its management. The best scenario according to the environmental categories of the LCA methodology was the scenario that integrates separate collection by waste streams, recycling of those materials, composting of organic waste and incineration of the residual waste.

Seeking for possibilities to embed CE at the country level, Cruz-Pastrana and Franco-García in Chap. 4 explored the feasibility to implement a cap-and-trade system (C&TS) to mitigate climate change effects which has been widely applied in other countries but not in Mexico. In such context, CE can offer additional possibilities to embrace C&TS generating both environmental and economic benefits. Hence the purpose of this chapter was to explore the C&TS feasibility in Mexico and to highlight the implications of considering CE models into a cap-and-trade instrument. Approaching this intention by using MACC (marginal abatement cost curves), 57% of all the measures to reduce GHG with negative cost agglomerates could contribute almost 1% to the GDP by 2020. Based on MACC calculations, the chapter argued that C&TS is potentially feasible in Mexico, even further as the attached carbon price by end-users might be a strong motivation to integrate CE tenets to their operations, bringing new innovative opportunities to close materials' and energy loops.

13.2 Stakeholders

Moving forward, the rest of the papers in this volume were introduced by following the stakeholder structure which was outlined in Chap. 1. We started with papers that clearly showed the public sector as one of the most influential stakeholders, in particular when it concerns the "household solid waste management" which is the focus of Chaps. 5 and 6. In the case of Chap. 7, a governmental programme to stimulate the urban symbiosis (recycling) implementation in China was the core of the analysis, whilst in Chap. 8 the Indonesian higher education management was the focus, education being one of the public sector's responsibilities. Universities were discussed as the intellectual cradle for the transition from linear to circular economy trends.

13.2.1 Stakeholder Promoter: Public Sector

It is not a coincidence that the authors (Suherman et al.) of Chap. 5 elaborated their case under the stakeholder identity and salience framework for their analysis of the household solid waste management in Cimahi City, Indonesia. In effect, the theory of stakeholder identity and salience by Mitchell et al. (1997) is the core analytical tool applied to this Indonesian case that to some extent reflects the current situation of the majority of cities in the country. Authors have a double interest whilst conducting this research; the first one is to approach systematically the identification and classification of the stakeholders involved in the household solid waste management in Cimahi City and secondly to detect the specific stakeholders who would be entitled to lead the transition towards circularity patterns of the solid waste management. Moreover, collaborative schemes for such transition can be discussed and designed by engaging those salient stakeholders. As one of the main findings of this research, it proved possible to identify and classify the stakeholders in Cimahi City for the household waste management into two meaningful categories: (i) the formal sector (government, NGOs and private ones) and the (ii) informal sector (not officially registered waste collectors and recycling entities).

Díaz and Tinoco-Castrejón presented in Chap. 6 the efforts of the government of Cuautlancingo, Puebla, Mexico, to transit the urban solid waste management towards circular economic practices. They interviewed governmental stakeholders to diagnose the current practices that resemble those of CE and validated such information through either secondary information sources or interviewing other stakeholders who were to some extent directly involved in the waste management. Based on the evidence, the authors showed that in spite of the municipal efforts to provide proper waste collection and disposal services, the urban solid waste management in Cuautlancingo is lagging behind when comparing it with what circular economy promotes. There is a consensus among stakeholders that the more discharged materials during production and post-consumption can be recycled within companies, a higher productivity level of the municipality can be achieved because this means a reduction of the materials to be treated or disposed by municipalities. The use of indicators to boost circular economy practices is suggested, such as the resource consumption rate (per capita), rates of segregated and recycled materials and rates of waste used for energy recovery. Additionally, local government is recommended to formulate regulations and policies in order to ensure the environmental care and responsible acting of the citizens. This should be integrated with awareness campaigns about waste management opportunities for the community. Openness for collaboration with other stakeholders was enthusiastically mentioned whilst talking to governmental officers.

Chapter 7, entitled "A Massive Urban Symbiosis: A Preliminary Review of the Urban Mining Pilot Bases Initiative in China", written by Xue et al. aims to bring the Chinese urban mining pilot base (UMPB) case under the light of the governance assessment tool (Bressers et al. 2016). The research question driving this work is what are the typical features as a "massive urban symbiosis policy" that the Chinese urban mining pilot base (UMPB) programme represents and to what extent can they be improved? The UMPB is a Chinese governmental initiative to stimulate recycling of key materials as the main urban symbiosis activity that links efficient material flows between production and consumption, the industry sector and the urban sector. The urban symbiosis (a concept derived from industrial symbiosis) is also in China framed under the CE policy and has been embedded into China's urban mining pilot base (UMPB) programme that promotes the growth of the recycling industry and the use and development of advanced technology to cope with high-value recycling practices. The current 45 pilot bases with a planned 6.6 billion tonnes per year capacity indicates that a massive urban symbiosis effort is taking place in China. Moreover, this paper shows that the driving force of waste recycling management in China is moving from an environmental and recycling angle to resource strategies. Despite UMPB's legal and subsidies assurance, this does not reflect the creation of a perfect governance context for its implementation. On the contrary, some incoherencies and policy conflicts were detected at the multi-ministerial cross-management network, who are the actual UMPB implementers. Those represent some of the main barriers (challenges) for this programme.

In Chap. 8, "Incorporating Circular Sustainability Principles in DKI Jakarta: Lessons Learned from Dutch Business Schools Management", Nurdiana et al. analysed the experiences of the integration of circular economy into the Dutch high education curricula with the purpose of using them as reference for the Indonesian higher education system. The authors identified the need to expand the concept of "green university" to the broader concept of CE. This is due to the important role universities play in societies, forming capacities to face current and future societal challenges, especially those oriented to cope with resource scarcity and pollution effects. As part of the findings in Indonesia, it can be mentioned here that the Indonesian universities have a relatively narrow view, whilst the Dutch cases analysed present a more holistic approach towards sustainability and CE, offering them some extra manoeuvre opportunities for the inclusion of CE education.

13.2.2 Stakeholder Promoter: Private Sector

The internationally operating supermarket chain Walmart in its Mexico operations has identified food waste as a niche for improvement as described in Chap. 9, where the concept of zero waste to landfill (ZWTL) was explored with the intention to ameliorate the organic waste management. ZWTL pursues prevention and reduction of wastes at their source of generation, which effects are in line with CE principles. In fact, Rincón-Moreno et al. suggest that "circularity" would increase productivity throughout the food supply chain if it became part of the business strategy. In order to test this, a showcase was selected to explore how the organic waste management can be improved by combining a CE business model and a ZWTL strategy. After analysing the data of the organic composition of the wastes (from the show case), the authors elaborated upon the idea that most of the food considered as waste can be recovered through different stages. Even further, 40% of the food waste management costs can be saved through three business actions associated with those recovery stages. The proposed framework intends to collect all the sustainable concepts that might potentially be resembled in other industries with similar challenges. The findings showed that most of the food which is considered waste (discharged organic products) can be recovered through different operations within the shop facilities: inventory management, first expired-first out, donations, livestock feed and a biodigester to generate electricity. The biodigester represents the most important investment, though 40% of the food waste management costs can be saved through three of the business actions associated to those recovery stages. Through the conduction of this project, a combined framework (SOL4FoodWaste) is elaborated and discussed.

The PetStar PET bottle-to-bottle recycling system, a zero-waste circular economy business model, was presented in Chap. 10. Cámara-Creixell and Scheel-Mayenberger emphasised the point of finding ways to organise the PET bottles post-consumption enabling PetStar to successfully contribute to the plastic recycling industry. Participation of three stakeholders involved in the industry – the environment, society and business – requires a different business model in which all actors can participate and produce a more inclusive added value. This is mentioned in reflection on the scavenger's conditions as an important stakeholder in the PET bottle recycling value chain. Scavengers mostly operate informally, resulting in their exclusion from traditional business models. At this regard, "PetStar is a company that has designed and implemented a Circular Economy business model for PET bottles that is trying to achieve a dream for the recycling industry, i.e. to disengage the recycled bottle from virgin resources, avoiding the conversion of the packaging to waste, and operate a perennial cycle in the use of the packaging". This case disclosed how their business model is operated in a sustainable way: economically feasible and competitive whilst environmentally resilient and socially oriented towards the poorest and most informal sectors, the scavenger and collector communities. Even beyond their own market boundaries, there is a strong belief that conditions can be created to enable replicability of their model of sustainable recycling to other developing countries where the potential of zero-waste circular value systems is well perceived by a large number of stakeholders in the PET bottle value chain.

In Chap. 11, the concepts of social life cycle assessment (S-LCA) and combined social and environmental LCA (SELCA) were explored through the application of existing LCA methods to the global value chain of denim jeans. Franco-García et al. reported on the SELCA method from this explorative research. The inclusion of the social component into the assessment of the denim jeans' showcase contributes to the battery of impact assessment tools for products whose value chain scope is multinational (global). Four scenarios for jeans assembly were compared; three of them are defined under the circular economy (CE) principles by including recycled materials (cotton, PET and nylon 6) during the yarn production. During the application of the SELCA method, some new challenges were encountered related to inventory analysis, in particular during data acquisition for social inventories. This last aspect is mainly due to the extensive list of key stakeholders for the showcase and the qualitative nature of the social metrics used. The list starts with cotton cultivators from different countries where regulations and codes of conduct seem to have contextualised interpretations and consequently different levels of implementation. In this regard, governmental intervention to instrument the transition towards suitable social/environmental performance along the global jeans value chain was also discussed in this paper.

13.2.3 Stakeholder Promoter: Civil Society

Plasencia-Vélez et al. presented Chap. 12 "A Circular Model of Residential Composting in Mexico State". This is the description of a composting project led by a civil society group within the physical borders of a residential unit nearby Mexico City. They produce compost from their own organic wastes, closing the loop by creating a product that nourish inner gardens of their residential area. The authors provided an in-depth description of their case arguing that their research contributes to the CE studies, in particular addressing the participation of households and, by this, showing that CE practices can be promoted by anyone who perceives the potential of discharged materials to be reintegrated into the natural biological cycle. The Mexican urban solid waste (USW) has a high organic component (45–55%) that can be potentially recovered and transformed as the households of Club de Golf Bellavista showed in this research. For the authors, this is a grass-roots example with positive effects on the USW management. Hence, with the intention to deploy its potential at a larger scale, the purpose of this research was to identify the necessary conditions for this type of projects to be successful. Therefore, it is important to identify and analyse similar cases, in order to learn from the existing best practices. Looking beyond the borders, British and Dutch cases are also analysed through an analytical framework constructed from specialised literature. It is used to assess the case in Mexico State by the means of surveys and semi-structured interviews. Among the most relevant findings, "social participation" came consistently across as a relevant success factor in this type of grass-roots initiatives. Additionally, promoters of this case gave suggestions about ways to engage the other neighbours in the separation phase of USW.

13.3 Concluding Overview and Suggestions for Future Research

The Ellen MacArthur Foundation (2015), SUN and McKinsey have estimated that "by adopting Circular Economy principles, Europe can create a net benefit of €1.8 trillion by 2030, or €0.9 trillion more than in the current linear development path". Certainly, it is much easier to calculate monetary benefits than to implement the necessary instruments to transit to a more circular economy through the current complexity of value chains. The transition needs to involve product designers, business developers, investors, marketers, managers, consumers and every single person of the society who has to deal with any type of discharged resources (waste) and can still see its value as a resource. More importantly, stakeholders are needed to bring the resources back to any of the closed material (resources) loops to avoid landfilling to its maximum expression (zero waste). This implies full systemic change and innovation at different levels: technological, organisational, social, financial and policy level. It is though expected that even in a highly CE scenario, there will remain some elements of linearity as virgin resources are required and ultimate residual materials are disposed of. Some technological challenges stem from the current waste treatment facilities, such as incinerators that do not fit within the CE principles. To deal with that some countries, e.g. the Netherlands, have considered a midterm plan to gradually transform low-efficient incinerators into specialised recycling and reuse facilities.

Also relevant is the growing movement of critical think-tanks to further develop new paradigms to practise the CE tenets. The number of approaches and business models to be considered grows constantly due to the sense of urgency produced by the foreseen beneficial potential of CE. Cramer (2014), for instance, has reported nine circularity schemes which at the same time imply the integration of related business models. Here is the list of her 9Rs: (i) refuse, prevent the use of resources; (ii) reduce, decrease the use of resources; (iii) reuse, find new product use (second hand); (iv) repair, maintain and repair; (v) refurbish, improve product; (vi) remanufacture, create new products from second hand; (vii) repurpose, reuse product for different purposes; (viii) recycle, reuse raw materials of product; and (ix) recover, recover energy from waste.

Historically, the practice of the 9Rs has been observed at a small and individual scale, but the real challenge is to scale them up into collaborative platforms where resource flows create value and close the production loops into a zero-waste system. The businesses that have invested in innovations like new material recovery methods and material substitution are expected to play a leading role on this because they have an influential position within their networks and stakeholders. On the other hand, the efforts of the private sector require the support of policy instruments and the consumer participation, as it was repeatedly mentioned in the cases here presented.

To sum up, the following reported barriers can be turned into the focus of further research along the ongoing CE implementation: (1) institutional barriers posed by existing business models, (2) organisational barriers as a result of poorly coordinated efforts, (3) legal barriers hampering innovation to some extent, (4) economic barriers that focus on current business models, (5) behavioural barriers to change consumers' attitudes, (6) environmental commitments associated to local and international agreements and (7) technological barriers that prevent incremental innovations.

With no doubt, technological innovations are a must in the transition towards CE, but it can also be argued that innovation is not always the panacea when the end-of-life of some materials represents a barrier by itself or not all the materials might be improved or substituted. Hence, in some cases, producers have to compromise the toxicity level of some ingredients in the formulation of their products, simply because they cannot sacrifice the product quality. In other cases, producers have important suppliers who cannot be replaced or do not agree to provide information on their materials. A way to overcome the reluctance to open up the flux of information could be by a suitable policy mix that regulates the transparency of information and/or stimulates purchasing circular products.

In regard to the local and international environmental agreements, e.g. greenhouse emission reduction (Paris 2015), CE is foreseen as a contributor to achieve the Paris' targets by including, for instance, a CO_2 tax per product which could result in a more realistic price of products and simultaneously have impact on Sustainable Development Goal 12 (responsible production and consumption). In reality, a plethora of environmental costs are not yet internalised in the product price, so this can be an opportunity to also reduce the prices of some products when they proved to be produced under any (or similar) of the 9R (Cramer 2014) frameworks.

The CE and ZW nexus deploys, as it was here above briefly discussed, a large number of current and future research opportunities. The chapters compiling this book have tried to illustrate such potential, but we, as editors, have to admit that the CE topic is such a prolific and dynamic field that results, currently, impossible to completely describe it. In other words, this book aims to be a piece of "work in progress" that connects two relevant topics to transform wastes into resources: CE and ZW.

References

Bressers H, Bressers N, Kuks S, Larrue C (2016) The governance assessment tool and its use. In: Bressers H, Bressers N, Larrue C (eds) Governance for drought resilience: land and water drought management in Europe. Springer International Publishing, Cham

Cramer J (2014) Milieu. Elementaire Deeltjes: 16. Amsterdam University Press B.V, Amsterdam

Ellen MacArthur Foundation (2015) Towards a circular economy. Ellen MacArthur Foundation, Brussels

Lansink A, de Vries H (2010) – in 't Veld, De Kracht van de Kringloop; Geschiedenis en Toekomst van de Ladder van Lansink (The power of closing the loop; history and future of the ladder of Lansink). Vrilan, Apeldoorn

Mitchell RK, Agle BR, Wood DJ (1997) Toward a theory of stakeholder identification and salience: defining the principle of who and what really counts. Acad Manag Rev 22(4):853–886

MIX
Papier aus verantwortungsvollen Quellen
Paper from responsible sources
FSC® C105338

If you have any concerns about our products,
you can contact us on
ProductSafety@springernature.com

In case Publisher is established outside the EU,
the EU authorized representative is:
Springer Nature Customer Service Center GmbH
Europaplatz 3, 69115 Heidelberg, Germany

Printed by Libri Plureos GmbH
in Hamburg, Germany